Expert Process Planning for Manufacturing

Expert Process Planning for Manufacturing

Tien-Chien Chang

Purdue University

Addison-Wesley Publishing Company

Reading, Massachusetts • Menlo Park, California • New York
Don Mills, Ontario • Wokingham, England • Amsterdam
Bonn • Sydney • Singapore • Tokyo • Madrid • San Juan

Library of Congress Cataloging-in-Publication Data

Chang, Tien-Chien, 1954-
Expert Process Planning for Manufacturing / by Tien-Chien Chang.
p. cm.
Includes bibliographical references.
ISBN 0-201-18297-1
1. Production planning—Data processing. 2. Expert systems (Computer science) I. Title.
TS176.C452 1990
658.5'00285—dc20 89-18341
CIP

ABCDEFGHIJ-MA-943210

To my parents Mr. and Mrs. Hsin Chang,
my wife Li-Fen,
and children David and Joseph

Preface

In the product realization process, the first and most important step is making a process plan. The quality of a product and the cost of producing it are strongly influenced by the process plan. Production planning, scheduling, part programming, facilities layout, etc.—all these functions take process plan as their input. In the past the majority of manufacturing systems were operated by humans. Since such a system responds slowly and is able to adapt to incomplete information, an inflexible and slow process plan generation mechanism is acceptable. Either manual process planning or retrieval-based variant process planning systems can satisfy the need. Today, the production method is gradually moving toward automation. Flexible automation has been especially stressed in recent years. The need for dynamic responses, fast plan generation, and smooth interface between design and manufacturing functions become essential in operating the new manufacturing systems. Thus, the automation of the planning function is critical.

Computer-Aided Process Planning (CAPP) or automated process planning is an approach that uses computers to generate a process plan. When constructed properly, such a system can satisfy the above mentioned needs. However, the task of automating the process planning function is not a simple one. No single algorithm can model the complexity of the thinking process of an experienced human planner. Thus, Artificial Intelligence (AI) seems to be a natural candidate for the application.

The development of CAPP started in the late sixties. The pioneers treated the process planning problem as a machining optimization problem. In the early seventies the Group Technology (GT) concept was introduced. Several GT-based retrieval systems were developed. *GT code, part family, standard process plan,* and *plan editing* were some terms familiar to users. Those systems by no means generated new process plans automatically. Yet, even the time and cost saved by using retrieval-based process planning systems pleased many users. Many of today's process planning systems are still of this type. This approach came to be called "variant" approach, since it varies the existing plan manually to make a new plan.

A system that can eliminate the manual editing work during process planning is desirable. In the mid-seventies the generative process planning approach was studied. Since then many other systems have been built. The majority of these systems either automate only a small portion of the overall process planning function or handle a very restricted domain of parts. Several such systems have been implemented in industry. All of them require certain degrees of human interaction. The eighties marks the beginning of AI in process planning. After years of frustration, a branch of AI—Expert System—finally paid off. Many successful implementations of expert system in industry have been reported. One of the fruitful application areas of expert system is in planning. The need in manufacturing process planning seems to match the expert system capability well. Many works have since been reported in the literature, yet few address the problem from an AI point of view and in an integrated manner.

The purpose of this book is to provide a complete, yet concise, introduction to expert process planning systems. Starting from the input or design representation, to the expert system formulation, every aspect of building process planning is discussed. Material collected from open literature and the research results developed by the author are arranged in a unified manner. This book also discusses research issues which need to be addressed. It can be used as a reference book for both researchers and industrial practitioners. This book can also be used as a textbook in a graduate level course on automated process planning. When complemented with reading assignments, it can cover a three-credit-hours course. Although not essential, knowledge about manufacturing processes, artificial intelligence, and computer programming will be helpful to readers. Other books published in this series can also be used to complement this book.

Since research in automated process planning has not reached its maturity, new developments occur at a rapid rate. It is not the author's intention to suggest that what is discussed in this book is the final word, or even the best approach. Rather, this book presents a summary of what has been done in the past decade in the field of automated process planning. The author hopes this book can provide new researchers guidance when they enter this interesting field, and can prevent redundant work being done.

As with any other book, the information has been collected from many sources. Much of the knowledge covered in this book is the result of the research work carried out by the author, his students, and colleagues. He especially would like to acknowledge the following: his former students Dr. Sanjay Joshi for his study on feature recognition, Mahesh Kanumury and Jatin Shah for their study on QTC process planning systems, James Moore for his work on cell control, and Professors Dave Anderson and O. Robert Mitchell for their cooperation in developing, respectively, the design and the vision inspection system for the QTC system. The QTC system is highlighted in Chapter 6 of this book as an example of an expert process planning system.

The author also would like to thank Mr. Chris Pochowicz for his editorial help. Many other individuals also contributed to the book in one way or another. I am grateful to the referees for their useful comments. The assistance and encouragement from series editors, Drs. David Dornfeld and Tony Woo are appreciated. Last but not the least, the author would like to thank Mr. Don Fowley, senior engineering editor at Addison-Wesley, for his patience and support throughout the project. It was a pleasure working with him.

The support from the Engineering Research Center at Purdue University, which was founded by the National Science Foundation, a National Science Foundation Presidential Young Investigator Award, and matching funds received from Rockwell International, Xerox Corporation, and Digital Equipment Corporation enabled the author to conduct the research which resulted in this book. Their kind support is deeply appreciated.

Contents

Expert Process Planning for Manufacturing

1

Introduction

This book discusses how to develop an expert process planning system. In a sense, a process plan can be considered a recipe. A recipe is "a set of instructions for making something from various ingredients."[1] Given a detailed recipe, a novice cook can usually prepare almost any meal. Of course, the quality of the meal still depends on the talents of the cook. Cooking is an art, and like all arts, a little magic "touch" is very important to the final result. A chef, then, is a master cook who normally does not require any written recipe. He or she knows how many ingredients to put into the dish, at what temperature to cook it, how long each step in the process takes, and just when to stir the food. Based on the chef's experience, a delicious dish can be prepared in no time at all. However, ordinary people, like most of us, are not as experienced. Without recipes, we are forced to order elaborate dishes from a restaurant. Who writes the recipes, then? Naturally, they are written by talented, experienced chefs. Knowing this, we can apply the same principles of the culinary art to manufacturing.

[1] The Merriam-Webster Dictionary, G. & C. Merriam Co., 1974.

ɡ

...at the domain of manufacturing, however, to discrete ...ng. In this kind of manufacturing environment, a master ... such as a tool maker, is equivalent to a master chef. Just ... what you want, and he can make it for you, with or without ...iled design drawing. Such a master knows how to run many ...es of machine tools. He can see in his mind the necessary processes a part needs. However, an ordinary machinist is not as talented. Without a detailed design drawing and manufacturing instructions, the process plan or recipe of manufacturing, he cannot complete the job. This process plan helps the machinist to complete the job step by step. When the manufacturing job is done by a group of operators, a process plan is essential so that the manager can schedule the jobs. In modern manufacturing, automation is a trend. Automated machines are not intelligent at all; they can only follow instructions. In order to run an automated manufacturing system, a detailed process plan (recipe) is a must, and the person who prepares process plans, a process planner, must be experienced in many aspects of manufacturing.

1.1 The Need for Process Planning

In the restaurant industry, there are fast food restaurants and restaurants for formal dining. In the manufacturing industry, there is mass production and batch production. A fast food restaurant begins with a few proven recipes, e.g., Big Mac, Quarter Pounder, French Fries, etc. The kitchen is designed to mass produce these products (foods). The manager of the restaurant can literally take someone from the street and put him or her to work in a few hours. World travelers will tell you that the Big Mac ordered from the McDonalds in Peking tastes exactly the same as the one from the McDonalds in their U.S. hometown. In manufacturing, mass production, production line, or transfer line works the same way. The line is designed to follow a proven process plan. Less experienced machine operators can be used to man the line. Just as fast food restaurants do not change their recipes often, mass production process plans are often not changed. What is important for such an environment is knowing how to optimize the process plan and the line design, so the system can run efficiently.

In a formal dinner restaurant, meals are made to order. In a very good restaurant, there may be a large menu from which to choose. Special instructions can be given to the kitchen to "customize" a dish. Here especially, the chef must be experienced and talented in order to correctly prepare such customized orders. In small batch manufacturing, the environment is very similar to those in the restaurant industry; however, there are some differences. Usually experienced machinists are needed to produce small batch manufacturing. Since machined parts are much more complicated than food dishes, however, one machinist may not know how to run all the machines needed for the part. A job shop may be designed to group machines which perform the same type of machining process in the same department. A part is routed through the job shop, and necessary operations are performed on the part. A process plan defines this route. Since a large variety of parts are made in the shop and new part designs arrive continuously, new process plans need to be prepared all the time. In order to reduce the level of skill required of the operators, the process plan needs to be extremely detailed. The most judgment is exercised at the process planning level, not at the operator level.

From this example, we can see that a process plan is a recipe for making manufactured products. It is essential in running a modern manufacturing system. Process planning can be defined as *an act of preparing detailed processing documentation for the manufacture of a piece part or assembly*. Depending on the production environment, it can be very rough as shown in Fig. 1.1, or it can be detailed as shown in Fig. 1.2. This detailed processing documentation may include processes selected, operation parameters, process sequence, setups, fixturing methods, and/or device programs (such as NC part programs). An expert process planning system denotes a computer software system which does process planning by using expert knowledge. We can consider an expert process planning system a subset of Computer-Aided Process Planning (CAPP) systems. Such a system can replace the human process planner. What is discussed in this book is actually more restrictive than the title may imply. The process planning domain is limited to the part being machined. In this book, we try to show the necessary information for developing an expert process planning system. The past research in this area is also presented.

Route Sheet	by: T.C. Chang
Part No. S1243 Part Name: Mounting Bracket	
workstation	Time(min)
1. Mtl Rm	
2. Mill02	5
3. Drl01	4
4. Insp	1

Figure 1.1 A rough process plan

PROCESS PLAN					ACE Inc.
Part No. S0125-F Part Name: Housing Original: S.D. Smart Date: 1/1/89 Checked: C.S. Good Date: 2/1/89			Material: steel 4340Si Changes: ______ Date: ______ Approved: T.C. Chang Date: 2/14/89		
No.	Operation Description	Workstation	Setup	Tool	Time (Min)
10	Mill bottom surface1	MILL01	see attach#1 for illustration	Face mill 6 teeth/4" dia	3 setup 5 machining
20	Mill top surface	MILL01	see attach#1	Face mill 6 teeth/4" dia	2 setup 6 machining
30	Drill 4 holes	DRL02	set on surface1	twist drill 1/2" dia 2" long	2 setup 3 machining

Figure 1.2 A detailed process plan

1.2 What Is Process Planning?

Process planning, as defined by Chang & Wysk [1985], is the act of preparing detailed operation instructions to transform an engineering design to a final part. The detailed plan contains the route, processes, process parameters, machines, and tools required for production. However, when used in different industries and different shops, the process planning functions may involve several or all of the following activities:

- selection of machining operations
- sequencing of machining operations
- selection of cutting tools
- selection of machine tools
- determining setup requirements
- calculations of cutting parameters
- tool path planning and generation of NC part programs
- design of jigs and fixtures

The degree of detail incorporated into a typical process plan usually varies from industry to industry. It depends on the type of

parts, production methods, and documentation needs. A
for a tool-room-type manufacturing environment typically
the experience of the machinist and does not have to be
any great detail. In fact, the instructions, "make as per part print,"
may even suffice. In the automobile industry and other typical mass-production-type industries, the process planning activity is embodied in the hard automation (the transfer and flow lines used for manufacturing component parts and assembly). For metal-forming-type manufacturing activities such as forging, stamping, die casting, sand casting, injection moulding, etc., the process planning requirements are embedded directly into the design of the die/mold used, where most process planning activity is fairly simple. Of course, making the die for the forming activities is a one-of-a-kind operation. The job-shop-type of manufacturing environment usually requires the most detailed process plans since the design of tools, jigs, fixtures, and manufacturing sequence, etc., are dictated directly by the process plan.

In order to perform the process planning activities, a process planner must

be able to understand and analyze part requirements,

have extensive knowledge of machine tools, cutting tools and their capabilities,

understand the interactions between the part, manufacturing, quality, and cost,

possess analytical capabilities.

The process planning activity has traditionally been experience-based and has been performed manually. A problem facing modern industry is the current lack of a skilled labor force to produce machined parts as has been done in the past. Manual process planning also has other problems. Variability among the planner's judgment and experience can lead to differences in the perception of what constitutes the optimal or best method of production. This manifests itself in the fact that most industries have several different process plans for the same part, which leads to inconsistent plans and additional paperwork. To alleviate this problem, a computer-aided approach is taken. Development in computer-aided process planning attempts to free the process planner from the planning process. Computer-aided process planning can eliminate many of the decisions required during planning. It has the following advantages:

1. It reduces the demand on the skilled planner.
2. It reduces the process planning time.
3. It reduces both process planning and manufacturing cost.

4. It creates consistent plans.
5. It produces accurate plans.
6. It increases productivity.

1.3 Approaches to Computer-Aided Process Planning

There are three basic approaches to computer-aided process planning—variant, generative, and automatic. The variant approach uses computer terminology to retrieve plans for similar components using table look-up procedures. The process planner then edits the plan to create a "variant" to suit the specific requirements of the component being planned. Creation and modification of standard plans are the process planner's responsibility. The generative approach, however, is based on generating a plan for each component without referring to existing plans. Generative-type systems are systems that perform many of the functions in a generative manner. The remaining functions are performed with the use of humans in the planning loop. Automated systems, on the other hand, completely eliminate humans from the planning process. In this approach, the computer is used in all aspects, from interpreting the design data to generating the final cutting path. The following sections provide a discussion of each of the approaches.

1.3.1 Variant Process Planning

The variant approach to process planning was the first approach used to computerize planning techniques. It is based on the concept that similar parts will have similar process plans. The computer can be used as a tool to assist in identifying similar plans, retrieving them, and editing the plans to suit the requirements for specific parts.

In order to implement variant process planning, GT-based part coding and classification is used as a foundation. Individual parts are coded based upon several characteristics and attributes. Part families are created of parts having sufficiently common attributes to group them into a family. This family formation is determined by analyzing the codes of the part spectrum. A standard plan consisting of a process plan to manufacture the entire family is created and stored for each part family. The development of a variant process planning system has two stages: the preparatory stage and the production stage.

During the preparatory stage, existing components are coded, classified, and later grouped into families (Fig. 1.3). The part family formation can be performed in several ways. Families can be formed based on geometric shapes or on process similarities. Several methods can be used to form these groupings. A simple approach would be to

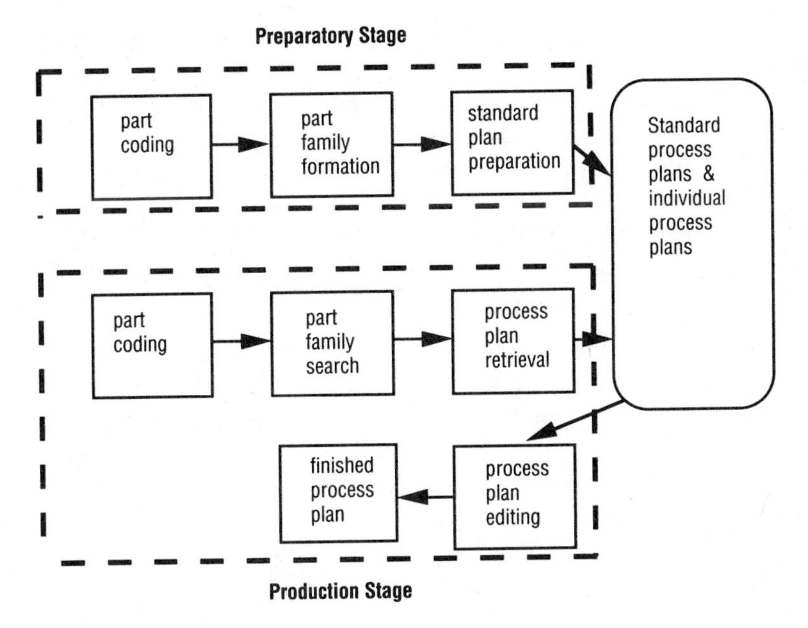

Figure 1.3 Variant process planning approach

compare the similarity of the part's code with other part codes. Since similar parts will have similar code characteristics, a logic which compares part of the code, or the entire code, can be used to determine similarity between parts.

Families can often be described by a set of family matrices. Each family has a binary matrix with a column for each digit in the code and a row for each value a code digit can have. A nonzero entry in the matrix indicates that the particular digit can have the value of that row, e.g., entry (3,2) equals one implies that a code x3xxx can be a member of the family. Since the processes of all family members are similar, a standard plan can be assigned to the family.

The standard plan is structured and stored in a coded manner using operation codes (OP-codes). An operation code represents a series of operations on one machine or workstation. For example, an OP-code DRL10 may represent the sequence center drill, change drill, drill hole, change to reamer, and ream hole. A series of OP-codes constitutes the representation of the standard process plan.

Before the system can be of any use, coding, classification, family formation, and standard plan preparation must be completed. The effectiveness and performance of the variant process planning system depends to a very large extent on the effort put forth at this stage. The preparatory stage is a very time consuming process.

The production stage occurs when the system is ready for production. New components can be planned in this stage. An incoming component is first coded. The code is then sent to a part family search routine to find the family to which it belongs. Since the standard plan is indexed by family number, the standard plan can be easily retrieved from the database. The standard plan is designed for the entire family rather than for a specific component, thus editing the plan is unavoidable.

Variant process planning systems are relatively easy to build. However, several problems are also associated with them. Some of these problems are the following:

1. The components to be planned are limited to similar components previously planned.
2. Experienced process planners are still required to modify the standard plan for the specific component.
3. Details of the plan cannot be generated.
4. Variant planning cannot be used in an entirely automated manufacturing system, without additional process planning.

Despite these problems, the variant approach is still an effective method; especially when the primary objective is to improve the current practice of process planning. In most batch manufacturing industries, where similar components are produced repetitively, a variant system can improve the planning efficiency dramatically. Some other advantages of variant process planning are the following:

1. Once a standard plan has been written, a variety of components can be planned.
2. Comparatively simple programming and installation (compared with generative systems) is required to implement a planning system.
3. The system is understandable, and the planner has control of the final plan.
4. It is easy to learn and easy to use.

The variant approach is the most popular approach in industry today. Most working systems are of this type, e.g., CAPP of CAM-I [Link 1976], Multiplan of OIR [1983], etc.

1.3.2 Generative Approach

Generative process planning is the second type of computer-aided process planning. It can be concisely defined as a system which automatically synthesizes a process plan for a new component. The generative approach envisions the creation of a process plan from information available in a manufacturing database without human intervention. Upon receiving the design model, the system is able to generate the required operations and operation sequence for the component.

Knowledge of manufacturing has to be captured and encoded into computer programs. By applying decision logic, a process planner's decision-making process can be imitated. Other planning functions such as machine selection, tool selection, process optimization, etc., can also be automated using generative planning techniques.

A generative process planning system comprises three main components:

1. part description
2. manufacturing databases
3. decision making logic and algorithms

The definition of generative process planning used in industry today is somewhat relaxed. Thus systems which contain some decision making capability on process selection are called generative systems. Some of the so-called generative systems use a decision tree to retrieve a standard plan. Generative process planning is regarded as more advanced than variant process planning. Ideally, a generative process planning system is a turn-key system with all the decision logic built in. Since this is still far from being realized, generative systems developed currently provide a wide range of capabilities and can at best be only described as semi-generative.

The generative process planning approach has the following advantages:

1. It generates consistent process plans rapidly;
2. New components can be planned as easily as existing components;
3. It has potential for integrating with an automated manufacturing facility to provide detailed control information.

Successful implementation of this approach requires the following key developments:

1. The logic of process planning must be identified and captured.
2. The part to be produced must be clearly and precisely defined in a computer-compatible format.
3. The captured logic of process planning and the part description data must be incorporated into a unified manufacturing database.

1.3.2.1 Part description methods for generative process planning systems

Part description forms a major part of the information needed for process planning. The way in which the part description is input to the process planning system has a direct affect on the degree of automation that can be achieved. Since the aim is to automate the system, the part description should be in a computer readable format. Traditionally, engineering drawings have been used to convey part descriptions and to communicate between design and manufacturing. Understanding the engineering drawing was a task suited for well-trained human beings and initially not suitable for direct input for process planning. The developments in CAD have provided some means for creating computerized storage of engineering drawings and for increasing feasibility for their use, as we shall see later. Several other methods have also been used for representing parts for generative process planning. Details are presented in Chapter 2. A few popular methods are presented here.

Codes

The first generation of generative process planning systems used GT codes or coding schemes based on other attributes to describe parts. The coding scheme consists of a sequence of symbols that identify a part's design characteristics and features, and/or its manufacturing attributes. Several commercial coding schemes are available, such as OPITZ [Opitz 1970], DCLASS [Allen 1979], and MULTICLASS [OIR 1983], KK3, etc. Variant process planning systems have been based exclusively on coding schemes. Several generative process planning systems such as APPAS [Wysk 1977], GENPLAN [Tulkoff 1981], COBAPP [Phillips 1978], M-GEPPS [Wang 1986], CORE-CAPP [Li et al. 1987], etc., have used coding schemes to describe parts for generative process planning systems.

Coding is typically a manual process and exact shape and size information, necessary for detailed planning, is lost when the part is described by a finite digit code. The degree of detail depends on the resolution allowed by the number and type of digits used. The code based representation is not suited for a completely automated process planning system, since the coding is a manual process, and a human interface is needed between the design and process planning function.

Special Descriptive Languages

The limitations in the use of GT for a completely automated system led to the development of special descriptive languages to assist in describing the part. The format of these languages allows planning to be performed easily from the information provided.

The AUTAP system [Eversheim and Esch 1983] used one such descriptive language. The part is described using both geometric and technological elements. Figure 1.4 shows an example of a part modeled using this approach. Besides using true geometrical elements, such as cylinders, cones, etc., it also uses subordinate elements such as patterns of holes. These elements are represented using key words

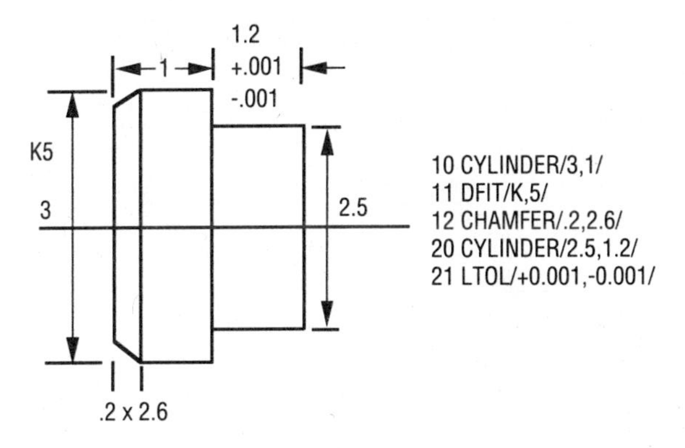

Figure 1.4 AUTAP part description

and attributes. The geometrical elements characterize the main contour, and the subordinate elements provide the details. Although reasonably complex elements can be modeled, this language lacks a complete set of Boolean operators and modeling a complex component may be difficult. Process planning is performed by analyzing the component description commands and relating them to manufacturing.

The CIMS/DEC part description system [Kakino et al. 1977] forms the input to the CIMS/PRO system [Iwata et al. 1980]. The part shape is described using volumetric elements obtained by revolving or parallel moving of the generating surfaces. The generating surface is composed of a concatenation of profile elements, which are given by directed line segments (Fig. 1.5). Technological information such as surface finish is associated and stored with each profile element. The goal here was to define a CAD/CAM oriented part description method that could be used for process planning. The problem is the limitation of this method to a limited domain of parts. Another similar approach based on syntactic methods was proposed by

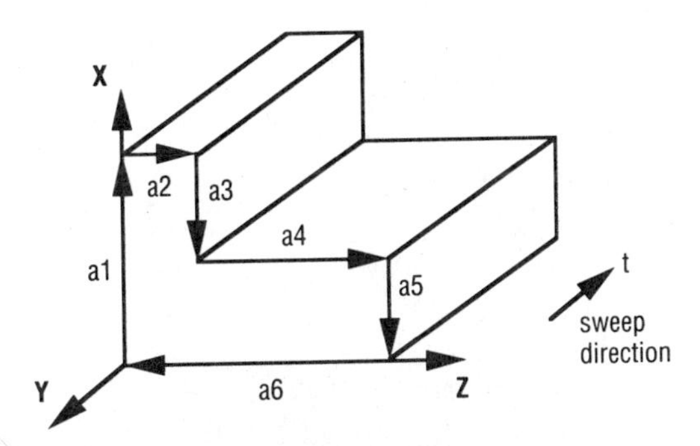

Figure 1.5 CIMS/PRO part description

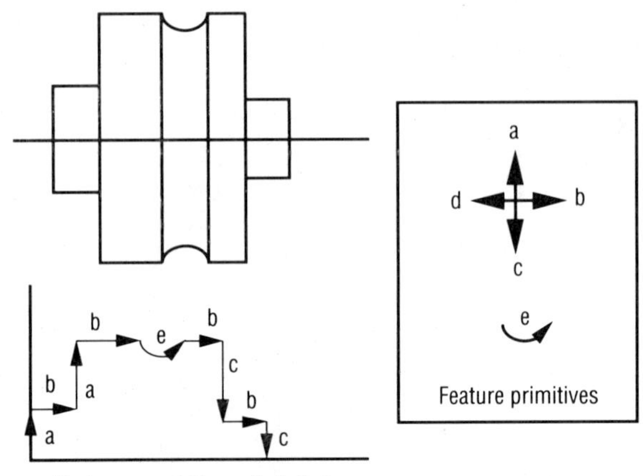

Part representation: **ababebcbc**

Figure 1.6 Syntactic part description

Jakubowski [1982] for describing parts. It uses the concept of basic primitives which are segments of straight lines or curves (Fig. 1.6). A contour describing the part boundary is constructed from the set of primitives described by a character string which is a left to right concatenation of primitives describing the part. This method is suited for part classification into families, by parsing the character strings describing the part. Since no technological elements are associated, it is not suitable for generative process planning.

GARI [Descotte and Latombe 1981] uses a part description based

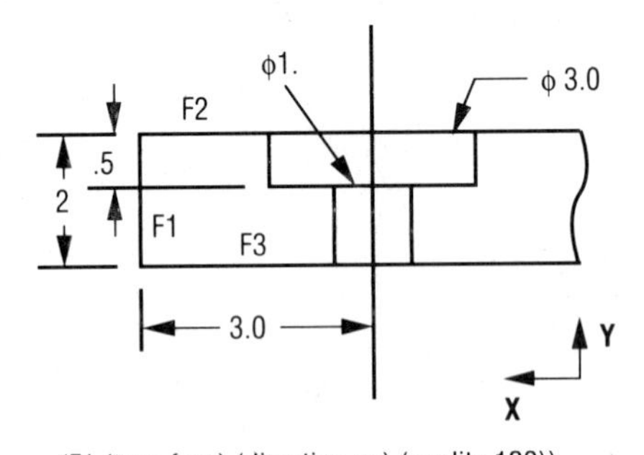


```
(F1 (type face) (direction xp) (quality 120))
(F2 (type face) (direction yp) (quality 64))
(F3 (type face) (direction ym) (quality rough))
(H1 (type countersunk-hole) (diameter 1.0)
    (countersik-diameter 3.0)
    (starting-from F2) (opening-into F3))
(distance H1 F1 3.0)
(countersink-depth F2 H1 0.5)
```

Figure 1.7 GARI part description

on a set of features, such as tapped holes, counter sink holes, bores, groves, notches, faces, etc., to describe the part (Fig. 1.7). These features are described using system words such as diameter, surface finish, distance, perpendicularity, tolerance, and so on. The relationships between features are also specified as attributes "starting_from" and "opening_into." The part description as a set of features conveys some expertise about its manufacturing when specified by this method. This representation is suited for planning tasks, but it has to be manually prepared for input into the system. For complex components this can be tedious and difficult.

The majority of generative process planning systems that use expert system approaches employ a special part description language—descriptive knowledge. They may appear in many different forms. Quite often too much information is embedded in the input. What is left for the process planning system is simple patter matching. While the use of a special descriptive language simplifies the process planning system, at the same time it requires more effort from the user of the system to prepare the input data.

CAD Models

Although the previous methods can provide information for several process planning functions, the part design still needs to be manually converted into one of these representations. Since a part can be modeled effectively using a CAD system, and the internal representation provides another computer compatible format, using CAD models as input to the process planning module has the potential of eliminating the human effort of translating a design into code or other descriptive form. One approach for converting the design model into a process planning system of usable information is through an interactive procedure [Chang 1982]. Although this approach still relies on human judgment, data handling is done by the system. It can be considered an aid in saving data preparation time and reducing data preparation error. Both 2-D and 3-D CAD models have been used and investigated for process planning applications. As defined earlier, those expert process planning systems which can reason directly from a CAD system are called automatic process planning systems.

1.3.2.2 Representation of process planning logic

Another major component of generative process planning systems is the process planning logic and knowledge. This includes the logic used by the process planner to make decisions on various aspects of planning, such as machine selection, tool selection, planning sequence, etc. Several methods have been used to represent this planning logic and data into formats and structures which will facilitate program coding and documentation. The decision logic has to be synthesized and structured in a clear manner, so that it is complete, unambiguous, nonredundant, and capable of expansion.

Decision Trees

A decision tree is a graph with a single root and branches emanating from the root. Each branch represents a possible course of action if the conditions specified to traverse the branch are true. Each branch can lead to another node or terminate in an action. The nodes provide further branching possibilities. When a branch is true, it can be traversed to reach the next node, and so on until a terminal point on the tree is reached. An example decision tree structure is shown in Fig. 1.8.

Decision Tables

A decision table is partitioned into conditions and actions and is represented in a tabular form. Decision rules are identified by columns in the entry part of the decision table. When all conditions in a decision table column are met, the marked decision is taken. A more detailed discussion on decision trees and tables can be found in [Chang & Wysk 1985]. The decision table has many advantages in structuring decision logic and can provide a nice modular structure, thereby simplifying maintenance and modification. An example decision table structure is shown in Fig. 1.9.

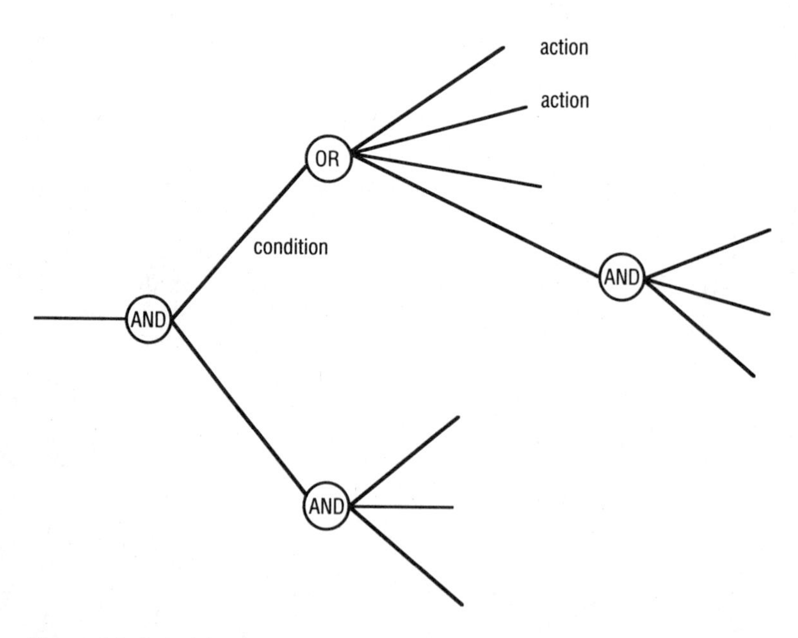

Figure 1.8 A decision tree

Condition	dia > 1.0	x	x				
	tol < 0.01		x				
		x					
			x				
			x				
Action	drill	1	1				
	bore		2				
	ream						

Figure 1.9 Decision table

Artificial Intelligence Based Approaches

The successful developments in AI/Expert systems have made them an important tool in the development of CAPP systems. The current use of AI in CAPP systems can be clearly divided into 2 parts:

1. use of AI for automated interpretation of the part description (see Chapter 4 for details)
2. expert systems for the development of the process plan itself (see Chapter 5 for details).

An expert system is *a tool which has the capability to understand problem specific knowledge and use the domain knowledge intelligently to suggest alternative paths of action*. Several aspects of the process planning problem make it amenable to solution by the expert systems approach. The problem has been traditionally experience-based, and it relies on much subjective and specialized knowledge acquired over long periods of time. Hence, systems attempting to automate process planning have to place a high priority on the reproduction, representation, and manipulation of the subjective knowledge as used by the process engineer. Expert systems provide an excellent framework to perform these tasks.

A typical expert system consists of (1) knowledge base of domain related facts, (2) rules for drawing inferences, and (3) inference mechanism for triggering rules, enforcing consistency, and resolution of conflicts.

Knowledge Representation

Several schemes have been developed for representing knowledge and rules in expert systems. These have been used extensively in the development of process planning systems:

1. predicate logic
2. production rules
3. semantic nets
4. frames
5. object oriented programming

Production Rules: This is one of the most commonly used knowledge representation schemes. Its basic concept is the notion of condition-action sets (or productions), and can be expressed simply in the form of IF-THEN rules.

IF [conditions] THEN [actions]

This approach has been used by several process planning systems, and its advantages and usefulness for encoding process planning knowledge is well documented. GARI [Descotte and Latombe 1981], TOM [Matsushima et al. 1982], PROPLAN [Mouleeswaran 1984], XCUT [Hummel and Brooks 1986], NBS's system [Brown and Ray 1987], and several others have used this approach. Rule-based systems have been developed to perform several of the process planning tasks, process selection [Nau and Chang 1983], tool selection [Gusti et al. 1986], fixture Design and set up [Englert and Wright 1988], process sequencing and operation planning [Barkocy and Zdeblick 1984], machining data selection [Wang and Wysk 1986], and CNC Milling [Priess & Kaplansky 1984].

Semantic nets attempt to describe the world in terms of objects and binary relations. According to a semantic net representation, knowledge is a collection of objects, and associations are represented as a labelled, directed arc. Semantic nets are easily understandable, but more difficult to implement. Figure 1.10 shows an example of a semantic net describing a hole.

Frames: A frame structure is a more general representation schema. It permits the representation of both procedural and declara-

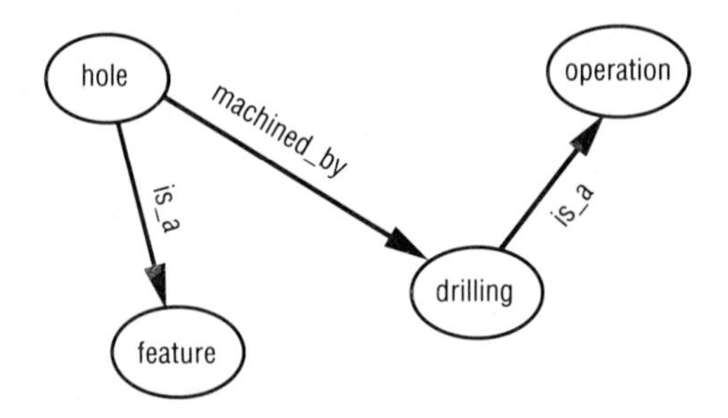

Figure 1.10 Semantic net representation

```
(
  FACE_MILLING
  (A_KIND_OF (VALUE SURF_BASIC_PROCESS)
  (FRAME_LABEL (VALUE FACE_MILLING))
  (TOOL_TYPE (VALUE (FACE_MILL)))
  (REQUIRED_MACHINE (VALUE (VERT_MILLING_MC HORT_MILLING_MC 5AXIS_MC)))
  (BATCH_QTY (VALUE 1))
  (ORDER_QTY (VALUE 1))
  (MACHINED_FEATURES (VALUE (FLAT_SURFACE STEP ISLAND)))
  (HARDNESS (VALUE 369))
  (WIDTH (VALUE (3.0 8.0))
  (SURFACE_FINISH (VALUE (126 249)))
  (DIMENSION_TOL (VALUE .01))
  (INDIVIDUAL_TOL (VALUE .005))
  (RELATED_TOL (VALUE .005))
  (PRE_MACHINED_FEATURE (VALUE FLAT_SURFACE))
  (PRE_SURFACE_FINISH (VALUE 700))
  (PRE_DIMENSION_TOL (VALUE .125))
  (PRE_INDIVIDUAL_TOL (VALUE .05))
  (PRE_RELATED_TOL (VALUE .05))
  (PRE_HARDNESS (VALUE SAME))
  (FINISH_ALLOWANCE (VALUE .08))
)
```

Figure 1.11 Frame for a face milling operation

tive information in terms of attributes, hierarchical relations with other frames, constraints, default values, procedures, etc. XCUT [Hummel and Brooks 1986] uses frame to represent the part knowledge. Several systems have been developed which use the notion of a frame to represent both the part and manufacturing knowledge, SIPPS [Nau & Chang 1985], SIPS [Nau & Gray 1986], and [Joshi, Vissa, and Chang 1988], etc. An example of a frame for representing an operation is shown in Fig. 1.11.

Object Oriented Programming: Object oriented programming is an artificial intelligence paradigm which provides an excellent data structure for symbolic manipulation of conceptual information. Most of the current concept of object oriented programming begins with the development the computer language SMALLTALK[Pinson and Wiener 1988]. With object oriented programming, details about objects can be ignored, and common properties could easily be inherited through property inheritance. Each object is an entity combining both procedural and declarative knowledge associated with some particular concept. The properties of these objects can be considered as attributes, entities, relationships, or procedural code. The concept of object oriented programming uses an idea similar to that of frames, but it provides a more flexible programming environment due to its use of data abstraction, inheritance, and modularity. Object oriented programming has been used by XCUT to represent the workpiece description [Hummel and Brooks 1986].

1.3.2.3 Databases and algorithms

Databases and algorithms form the final major component of a generative process planning system. Several databases are required to provide the process planning system with the information required to make decisions. Examples of the various databases required are shown in Fig. 1.12. In any given system, usually not all databases are present; the few that are used depend on the final output desired from the system. The databases usually contain company specific information and have to be specifically tailored to each company.

Algorithms are used primarily to perform the computations and guide the system during the decision making process. Typical applications of algorithms include determining optimum cutting conditions, such as machining and force requirements, tolerance distribution, process parameters, such as speed, depth of cut, feed, etc., which are subject to solution by means of empirical formulas.

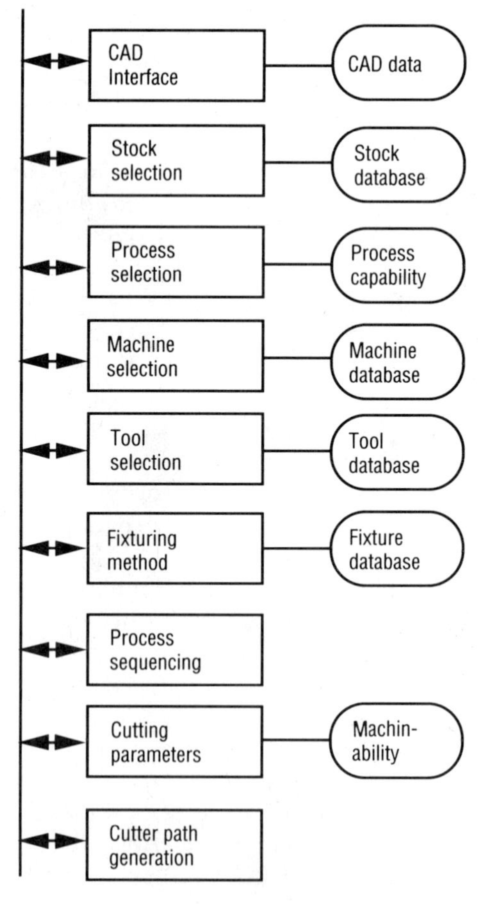

Figure 1.12 Process planning functions and supporting databases

1.3.3 Automatic Process Planning

The third approach is called automatic process planning. It denotes process planning that can generate a complete process plan directly from an engineering design model (CAD data). There is no human decision making necessary. An automatic process planning systems possesses two special features: the first is an automated CAD interface; the second is a complete and intelligent process planner. Currently, there are three major schools of thought as to the kind of CAD data to use as the input. One approach is to take a general CAD model (in order to provide complete and unambiguous data, solid models are used) and develop an interface to recognize the manufacturing features from this model. This approach has been taken by Choi, et al. [1984], Henderson [1984], Joshi and Chang [1988], etc. The advantage here is that general modeler can be used for design. However, recognition is a serious problem, as it is extremely hard to recognize complex features. More details about this approach are presented in Chapter 3 of this book. It is also worth noting that even the above referenced studies are not integrated. They study only the CAD interface.

The second approach uses a specially designed CAD model incorporating shapes that are immediately recognizable by the manufacturing planner. These familiar shapes, or features, are designed around manufacturing operations. Such a feature-based design environment limits the designer to the available manufacturing features, which are by definition feasible for manufacturing. By using a feature model, one can ensure that the parts designed can be manufactured. Macros for each feature can be written to synthesize operation plans and part programs. The feature-based approach has been taken by many generative systems, such as, AUTAP [Eversheim, et al. 1980], APPAS [Wysk 1977], CADCAM & TIPPS [Chang and Wysk 1985], XCUT [Hummel and Brooks 1986], [Kramer and Jun 1986], First Cut [Cutkosky et al. 1988], etc. The major shortcoming of the approach is that the manufacturing planning job is moved to the design stage. Another limitation of the approach is that features used in the design must be functional shapes for the design task, instead of manufacturing shapes. Most of the systems mentioned in this paragraph are not intelligent enough to be considered automatic.

The third approach is a hybrid approach; dual models are used. Pointers are used to link the two models. A design can be done using a feature-based modeler. The feature model is then evaluated into boundary models. Since geometric reasoning can be performed on the boundary model, it reduces constraints a designer has to obey during the design stage. The geometric reasoning is made easier because the design features are known to the system. QTC, the first system to take this approach, is presented in Chapter 6.

As you can see, the major difference between an automatic approach and other approaches is the automatic CAD interface capability. Of course, the system also needs to be intelligent. The planner reasons with not only the technological information, but also the geometric information. Since complete technological and geometric information are available to the planner, and the planner is able to reason with them, more planning functions can be handled. As a result, an automatic process planning system tends to be more complete, functionwise.

1.4 Historical Background of CAPP Development

In this section, the important historical events which led to the CAPP development are briefly reviewed. Since CAPP is a bridge between the design and the manufacturing functions, it is appropriate to discuss the significant development in both areas. The development in Computer-Aided Design (CAD) and Computer-Aided Manufacturing (CAM) is depicted in Fig. 1.13. CAPP is part of the CAM functions. The development of Computer-Aided Manufacturing can be traced back to the late 1940s when a numerical control concept was first proposed. The development of an NC machine tool marks the beginning of the CAM hardware development. After the first NC machine was built, it was realized that NC part program preparation was extremely tedious if done by hand. As a result, the development of a part programming language (APT) was started in the mid-1950s. The APT language development dealt with the data processing aspect of the CAM; it supported the operation of NC machines. Initially the APT language handled only simple geometries, such as point, line, circle, plane, quadratic surface, etc.

In the later part of the 1950s, NC machines became commercially available. New devices such as tool changers and new machines such as machining centers were developed. By the 1960s, several thousand NC machines had been installed in industry. NC control was put on nearly every kind of machine tool. Those machines replaced hand skill with programming skill. Less skilled machinists can now produce much more complex parts at a much faster rate and with more consistent outcome. The industrial robot was also invented during this period. It could replace some simple material handling tasks which used to be handled by human operators. Since early industrial robots were not computer controlled, the integration of robots with machine tools was not possible. As you can see, the major advancement on the CAM side, before the late 1960s, was the development of manufacturing hardware automation. The development was slow due to the lack of powerful computers and software tools.

On the software side of the development, Artificial Intelligence (AI) research started in the mid-1950s. A LISP language was invented

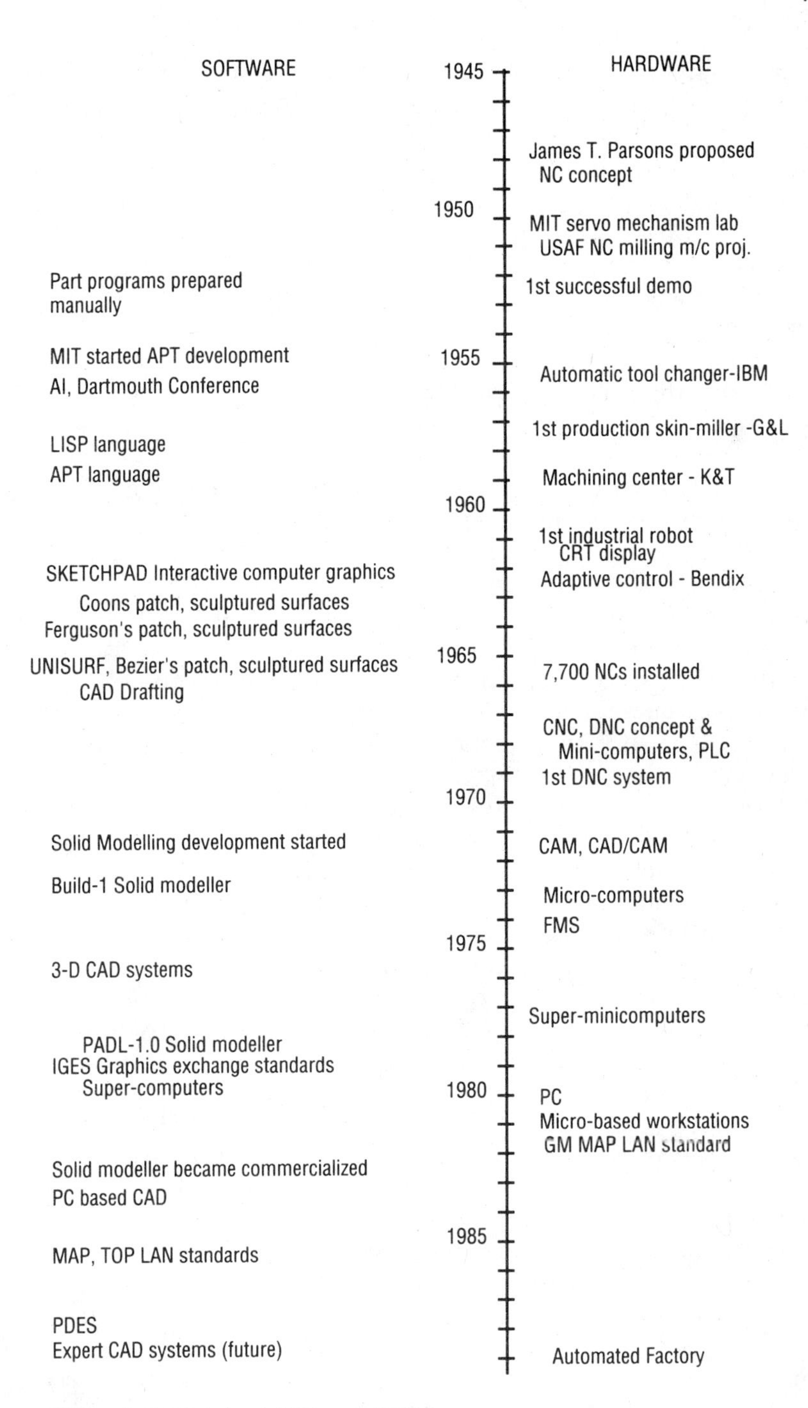

Figure 1.13 History of CAD and CAM

during that time. People had high expectations for AI. Due to oversell of the technology and lack of both software and hardware support, the AI rush soon became history. Not until the mid-70s did interest in AI pick up again. In the early 1960s, the invention of interactive computer graphics and sculptured surface models resulted in the development of CAD drafting and surface modelling systems.

By the late 1960s, mini-computers became available. Although they were slower than the mainframes, they were cheap enough that small engineering departments could afford them. CNC (Computer Numerical Control) machines were developed soon after. CNC is more flexible than hardwired NC machines. They use a general purpose computer as the center of the controller instead of dedicated electronic circuits. A CNC controller also provides communication capability that enables the machine to be more easily interfaced with a computer. When linking a central computer with several CNC machines, a DNC (Direct Numerical Control) is formed. The DNC system concept was soon turned into FMS (Flexible Manufacturing System). An FMS is a DNC system with automated material handling subsystems. The first FMS system actually used a mainframe computer (IBM 360) as the system controller. A powered loop roller conveyor was used as the material transport device. Parts mounted in a fixture, which in turn is mounted on a universal pallet, travel on the conveyor to the machining centers around the conveyor. FMS system control software utilizes the schedule generated by an offline software simulation model to control the system operation. Although it was hoped that 90% system efficiency would be achieved. In reality, only 70% can be achieved. The lack of operation support, such as scheduling, material and tool management, process planning, and part programming, are some of the problems. Therefore, although a delpha study conducted in the early 1970s predicted a rapid growth of FMS (thousands by the mid-1980s), there are only about 50 FMSs operating in the United States today.

The other significant development in CAM during that period was the invention of Programmable Logic Controller (PLC). PLC was initially designed as a replacement of relay panels for industrial control. PLC is a computer-based device. It allows a user to key in a control ladder diagram instead of doing the hard wiring. It eventually became the most important control device on the shop floor. The development of mini-computers also affected the robot control. Robots controlled by a mini-computer can perform much more complex maneuvers than those powered by peg boards or sequence drums. With better computers and sensors, industrial robots began to see and to feel. They became faster and more accurate. The challenge became how to make them easily integrated into a manufacturing system and how to make them easier to program. The concept of Computer-Aided Process Planning was proposed during this period.

The 1970s were also a time for micro-computer development. Many of the industrial controllers employ single board micro-computers. Applications such as CNC control were gradually replaced by micro-computers. The development of real time Operating Systems (OS) for the minis and simple Disk Operating Systems (DOS) for the micros enable more manufacturing applications to be computerized. Computer-based control devices became available and affordable. The development of super-minicomputers in the late 1970s enabled many engineering applications to be done locally in the engineering department. CAM became more affordable. The development of super-computers further provided computational power to scientists and engineers. On the software side, CAD drafting and CAD-based engineering analysis became mature. Solid modelers were also developed during this time, although they are not mature enough to be used in real design applications. The data exchange problem due to the success of installations of many different CAD systems prompted the development of IGES graphics exchange standards in the late 1970s. Some commercial CAPP systems became available during the later part of the 1970s.

By the 1980s, the micro-computer became increasingly popular. By the mid 1980s, there were millions of personal computers installed in offices, homes, and shop floors. Simple machine control can be done by personal computers. Engineering workstations were developed in the early 1980s and soon became popular in universities labs and industries. Many CAD and CAM applications that used to run on super-minicomputers can now run on desktop engineering workstations. Another significant development was Local Area Network (LAN). LAN allows many computers to be networked together to share resources such as disks and printers. The GM MAP LAN standard was proposed to standardize LAN protocol for shop floor application. On the software side solid modelers were finally used by designers. Many vendors started to deliver solid modelers. Since the speed of an average personal computer increased from the less than 0.5 MIPS (Million Instructions Per Second) in the early 1980s to the 1 to 5 MIPS at the end of 1980s, the memory increase from less than 64K byte to several M byte, and the disk space from less than 1 M byte to 20 to 80 M byte, the CAD and CAM software can now run on personal computers at a reasonable speed. Engineering workstations by the end of 1980s can run at more than 10 MIPS, have a several hundred M byte disk, and provide high resolution bit-mapped color graphics. Ethernet LAN and UNIX operating systems also became the standard. The computational speed is no longer a concern for many CAD and CAM applications.

In the early 1980s, several highly successful Expert System (ES) applications, such as the VAX computer configuration system and XCON of the Digital Equipment Corporation, resurrected AI. Subsequently, many manufacturing applications have been developed, including many ES based process planning system prototypes. Increasing interest has been placed on using ES for manufacturing planning, manufacturing system design and control, machine and system diagnosis, etc. There is a renewed hope of using computers to replace and/or assist humans in making decisions. Industrial automation is moving gradually from hardware to software. Whether the dream of having a thinking computer will ever come true is still questionable. Definitely, many of the simpler human decision-making tasks can be and have been automated. The 1990s will prove to be a time of information automation.

We have completed a brief discussion of the history of CAD and CAM development; now it is time to look a little closer at the development of CAPP. The idea of using computers in process planning activity was discussed by Niebel [1965]. Other early investigations on the feasibility of automated process planning can be found in Sheck [1966] and Berra and Barash [1968]. Many industries also started research efforts in this direction in the late 1960s and early 1970s. Early attempts to create automated planning systems consisted of building computer-assisted systems for report generation, storage, and retrieval. When used effectively, these systems can save up to 40% of a process planner's time. A typical example is Lockheed's CAP system [Tulkoff 1981]. Such a system can by no means eliminate the process planning tasks; rather it helps to reduce the clerical work of the process planner.

Most CAPP systems developed in the 1970s used GT retrieval as the major approach. In the 1980s more generative and knowledge based approaches were used in CAPP development. Two small scale surveys were conducted by the author in 1986 and 1988 [Haas and Chang 1987, Ahlgrim and Chang 1989]. The results are shown in the following. The types of products planned on CAPP systems in industry can be seen in Table 1.1. The percent in the parentheses shows the percent of systems among those replied has that feature. The capability of those systems in terms of output information generated is shown in Table 1.2. Table 1.3 shows the approaches taken by current systems. The input a system takes is summarized in Table 1.4. It's worth noting that the capability does not necessarily mean the particular function is done automatically. Quite often the function is done interactively. The 1986 survey is based on 41 companies' replies which use CAPP out of 196 questionnaires distributed. The 1988 survey is based on 31 companies which use CAPP out of 224 questionnaries distributed. The survey was sent to companies which had reported the use of CAPP and also those aerospace and machinery manufacturers. For out of the United States, only those who reported their CAPP work in open literature were surveyed. There is no indication as to the percentage of industry which uses CAPP.

Table 1.1 Types of products planned on CAPP system

Product	1986 No.	Percent	No.	1988 Percent
Turned Parts	25	(81%)	21	(75%)
Prismatic Parts	21	(68%)	17	(61%)
Sheet Metal	19	(61%)	11	(39%)
Bulkforming	10	(32%)	5	(18%)
Electronic Assembly	13	(42%)	8	(29%)
Mechanical Assembly	—	—	7	(25%)
Others	12	(39%)	4	(14%)

Table 1.2 Output Information Generated

Function	1986 No.	Percent	No.	1988 Percent
Processing Sequencing	30	(97%)	28	(100%)
Stock Selection	16	(52%)	10	(36%)
Machine Selection	24	(77%)	25	(89%)
Tool Selection	21	(19%)	17	(61%)
Machining Parameters	20	(65%)	14	(50%)
Cutter Path	10	(32%)	6	(21%)

Table 1.3 System Approach

Function	1986 No.	Percent	No.	1988 Percent
Variant	4	(14%)	4	(14%)
Generative	16	(52%)	6	(21%)
Knowledge-based	24	(77%)	3	(11%)
V-G Hybrid	21	(19%)	6	(21%)
G-K Hybrid	20	(65%)	2	(7%)
V-G-K Hybrid	10	(32%)	5	(18%)
Others	00	00	2	(7%)

Table 1.4 Input Information

Input	1986 No.	Percent	No.	1988 Percent
GT Code	4	(14%)	6	(23%)
CAD database	16	(52%)	6	(23%)
B-rep/Solid model	24	(77%)	3	(12%)
Raw Dimensional data	21	(19%)	6	(23%)
Relational Database	20	(65%)	2	(8%)
Special format	10	(32%)	4	(15%)
Others	00	00	2	(8%)

From the above data, one can conclude that the majority of current systems are designed to handle either turned parts or prismatic parts. Most of the systems output process sequencing and machine selection. The approaches—generative, variant-generative, and variant-generative-knowledge-based, and variant—are used nearly equally. Few industrial systems use an expert process planning approach. As for the input data format, few systems have interface with solid modellers. This is partially due to the fact that solid modellers are not popular in today's industrial design department, and partially due to the inability of current systems to deal with geometric reasoning problems.

1.5 Future Trend of CAPP

As it has been presented, the development of Computer-Aided Process Planning has been around for more than two decades. At the beginning, the approach was to find optimum machining parameters and cut distribution. Then, the approach evolved into report generation and documentation retrieval. In the later case, GT was used to help locate similar parts, thus becoming process plans. It was not until after ten years had elapsed that some kind of generative approach was taken. Although the introduction of AI and an expert system boosted both the interest in the problem and the capability of the systems, the results are still far from desirable. A majority of CAPP systems are still computer *aided*, instead of computer automated. Since an expert system enables nonmanufacturing experts (knowledge engineers) to build a prototype process planning system quickly, many people choose to do just that. Many of these systems contain a handful of imaginary manufacturing rules, and may require more effort to prepare the data input than would be required to complete the process plan. The idea that knowledge can be acquired from the interviewing of machinists by knowledge engineers hindered progress in building truly intelligent systems. In order to achieve our goal of building automatic process plan systems, a scientific study must be conducted on the manufacturing knowledge base. A mixture of science-based knowledge and experience-based knowledge is crucial. Whenever science-based knowledge can be obtained, it should be used instead of experience-based knowledge.

Figure 1.14 shows some of the approaches currently taken and their capabilities. The major differences among them are the CAD input format and the degree of automation in the planner. When the input is a 2-D drafting, the automatic drawing interpretation is almost impossible in a general sense. Although there were a few attempts, their domain is very restrictive. The majority of systems which take 2-D drafting as input use a human drawing interpretation and a variant process planning approach. For generative process planning, a 3-D CAD model that provides more information is more likely to be used. A feature-based model is a special kind of solid model. The

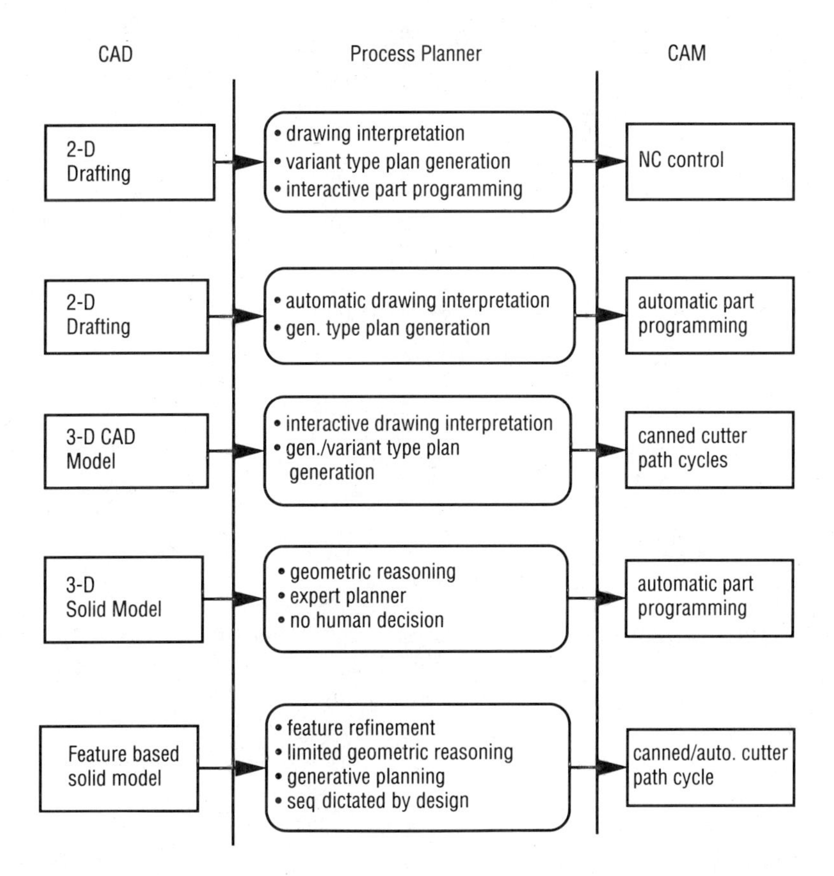

Figure 1.14 Some process planning approaches

difference is that it carries additional semantic information. The geometric reasoning is made easier because of the semantic information embedded in the feature models. For research, the current trend is to use either a regular 3-D solid model or a feature-based solid model as input.

When we look at the development of CAPP, we can see that it is a relatively new technology. From Fig. 1.15, we see some major advancement in CAPP. The figure shows the time versus the intelligence of the CAPP system. In the figure we use the intelligence of the system as a measure of the capability of the system. The higher the intelligence measurement, the more sophisticated the system is, and the better it can generate plans. Eventually, we will be able to build process planning systems which will have the same intelligence level as a human planner. However, this intelligence level is specific to the process planning task that includes design interpretation, detailed process plan generation, fixture planning, process economics analysis, and NC cutter path generation.

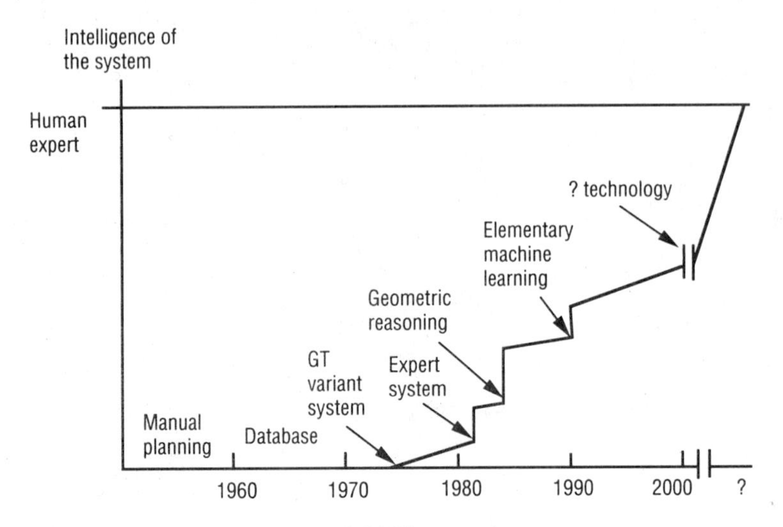

Figure 1.15 The development of CAPP systems

As you can see in Fig. 1.15, before the 1960s only manual process planning was used. In the 1960s data base was introduced to store and to help format process plans on a computer system. Since those systems do not make help in making decisions, there was no intelligence built in-to the systems. The decision is made by the user of the system. In the mid-1970s, GT was introduced to build variant process planning systems. Again, the plan creation and modification were done by the human user. However, as systems evolved, some intelligence was included. For example, some systems were able to find optimal cutting parameters, estimate time and cost, etc. Some decision tree and decision table based generative process planning systems were also developed.

In the 1980s, two major developments started. The application of an AI expert system in building process planning systems enables us to better model the process planning knowledge. The second development is geometric reasoning for process planning. Geometric reasoning (either feature recognition or feature refinement) allows the system to understand the design model, thus making more intelligent decisions. Due to these two technologies, process planning systems can perform closer to a human planner's level. However, they still need much development before they can achieve the level of human intelligence. We believe that in the early 1990s machine learning technology will have another major impact on process planning systems. The methodology will continue to improve, thus increasing the intelligence of the system. We believe that in order to truly reach the level of a human process planner, a breakthrough in computer technology, both hardware and software, is needed. Further study in the science of processes and the understanding of how the human brain functions will also be necessary to make this happen. We don't know when it will happen, but it is a goal of many of the researchers in various fields.

For the short term (five to ten years), we believe the major development will center around the integration of the process planning system with other design, manufacturing, and business functions. The availability of a common database, fully shared between design, manufacturing, planning, scheduling, control, and purchasing, will help in providing the completely integrated CAD/CAM environment capable of on-line generation of process plans. Process plans will also be generated based on shop floor scheduling and planning information. It will be used globally as opposed to the current local use. The process plan can be generated in real time when it is needed.

To be more specific, a list of research and development agenda follows:

a. CAD interface—both feature recognition and feature refinement.

b. Knowledge acquisition—how to help manufacturing engineer/knowledge engineer capture the ever changing shop and process knowledge. Machine learning issue also needs to be addressed in acquiring new knowledge.

c. Develop better CAD systems which provide complete, unambiguous information of the part on both geometry, topology, and technological information.

d. Better understanding of the science of manufacturing processes.

e. Develop interface with shop floor control, scheduling, purchasing, marketing, and accounting systems.

f. Consider problems in other product domains, such as metal forming, composite manufacturing, mechanical and electronic assembly, etc.

Future process planning systems will be used by the designer as a design evaluation tool and by the marketing department as a quotation calculation tool. They will also be used by the production control department as input to the master scheduling system and by the shop floor control and scheduler for the real time scheduling input. In order to do all of the above mentioned functions, the system will need real time data from all related departments. We will find the future process planning system to be an integral part of the overall Computer Integrated Manufacturing (CIM) system. It will be a vital link between the design and the manufacturing systems. This book does not attempt to cover how the future process planning system should be built. This is left to the researchers to decide. This book tries to show what has been done and the pros and cons of each method.

1.6 Expert Process Planning Systems for Industry

As can be seen from the previous section, CAPP systems have been developed and used in a few limited application domains. Most of the systems used in industry today are still of a variant or semi-generative nature. Great savings have been reported by industry who used even variant process planning systems. However, as we know that a variant process planning system does not help in making decisions. The system simply helps maintain a large database and provides the means to retrieve any previous written process plans based on GT. The planning decision still must be made by experienced human planners. The benefits stem from the standardization of existing plan reference, and easy editing capabilities. On the other hand, based on a study done in early 1980s, there was a serious shortage of process planners in U.S. industry. We have not seen any infusion of large numbers of process planners in the past several years. Realistically, then, we can conclude that the shortage problem still exists. What relieved the problem somewhat is the fact that the U.S. manufacturing industry has constantly declined. In order to gain back the competition edge and survive in the future world economy, we will need to enhance our manufacturing capability. To increase the manufacturing activity or even just to maintain it, the process planning capability needs to be enhanced. Given the reduced number of available experienced human planners and the fact that a process planner takes a long time to train, a more advanced process planning approach is essential.

It is well known that the only way to survive the future competition in the market place is flexible manufacturing. Flexible manufacturing here means quick adaptation to product and the demand changes. In such a manufacturing environment, shorter production leadtime must be realized. In order to reduce the production leadtime, the process planning time must be reduced. However, planning by humans is a stochastic process. Unlike a machine, humans tend not to work on something continuously. Many interruptions, caused either by the environment or by the planner himself or herself, are the major cause of delay. It is always desirable to remove this unpredictable time element.

When selecting new technologies to solve their problems, people in industry are always conservative. They tend to use only proven technology. The low tech approach of variant process planning is proven safe. Since the knowledge base still resides in a human planner's head, there is little concern about whether the system is generating the correct plan. The safeguard is the human planner. It has been shown that most people in industry don't even trust generative process planning [Section 4.4, Nolen 1989], let alone expert process planning systems. When the part domain is limited to one or a few families of parts with distinctive features, a generative process planning approach is superior to a variant process planning approach. In reality, many production environments are of this type, i.e. parts used in different models of the same product line.

An expert system tool can be used in a generative process planning approach to make a more powerful system. It can also provide more powerful expressive power to model the knowledge. Those features allow us to create a more automated and better process planning system. However, there are several problems that hindered the acceptance of the expert system tool by industry. It was legitimate to consider these as problems just a few years ago. As the computer technology advaces faster than one can imagine, these problems are diminishing. Following, we will discuss these problems or myths.

a. It is difficult to gain the expert knowledge.

Knowledge is crucial to the expert system. Knowledge acquisition from human experts is always difficult. As we experienced more about building expert systems, more methodologies have been developed to guide and assist expert system builders to acquire domain knowledge. Usually it takes years for a human planner to acquire his or her knowledge. This process has to be repeated for every human planner we hire. After knowledge has been acquired and properly represented, duplicating knowledge in expert systems is much easier.

b. An expert system is slow.

An expert system is usually slow. In order to come up with an answer, an expert system has to do a lot of search, pattern match, etc. This process is not very efficient as far as computation is concerned. Just a few years ago, expert systems had to be run on special designed Lisp machines in order to have a reasonable response time. Thanks to the advances in computer technology, we now have inexpensive machines which run at the speed of mainframes ten years ago. The speed of the computer in which the expert system runs has drastically improved; thus, the expert system performance has improved. Unlike in many real time applications, process planning does not need to be done in a split second. Minutes are acceptable time units in process planning. Today's expert system is definitely fast enough to provide the process planning needs.

c. Building an expert system is expensive.

Building an expert system is a very time consuming task. It also requires an expensive computing facility and software tools. The manpower part of building an expert process planning system is not going to be cheaper. However, as the computing facility and software tools become cheaper, the cost of building and running expert systems comes down with them. Just a few years ago it was not justifiable to run a small expert system application on a dedicated computer worth more than a hundred thousand dollars. Today, the same application may be run on a personal computer costing less than ten thousand dollars. At the same time, this personal computer may be shared by other applications as well.

d. An expert system is incomprehensible to an ordinary person.

People always don't feel comfortable to relay on a black box. Ten years ago, it was claimed by many AI experts that there were less than 100 people in the United States who were qualified to build an expert system. Today, expert system technology is no longer a myth. Although it is still not as straightforward as a procedural language program, it is not as cryptic as it was once described. Many engineers have been trained to use expert system tools. They will be able to carry out the development of and use expert process planning systems.

Expert systems have been successfully used in a large number of other industrial applications. Skepticism has disappeared after many successful systems have been demonstrated. Planning has been one of the main application areas of the expert system. The expert system is the most promising tool to solve the automated process planning problem. Since the introduction of expert system technology to process planning is relatively new, few, if any, expert process planning systems have been implemented in industry for routine use. However, several prototype systems have been built or are under development by industry (XPS-E [LaTombe, et al. 1984, 1985], Intelligent Machining Workstation [Cincinnati Milacron 1989], etc.). Although the technology has not reached a mature stage, it can definitely benefit industry in problems of a smaller domain. The study of expert process planning systems should not be left to researchers only, developers and industrial users should also try to harvest the fruit.

Quoting a call for a proposal from CAM-I, the industry sponsored international research organization which is credited with the development of the well known "CAPP" system, "The PPP (Process Planning Program) first began investigating AI-based process planning in the early 1980s, with the design of the XPS-E system. This system was initially intended to implement a modest program using current expert system technology and experience gained from the GARI project. Unfortunately, expert system technology was in its infancy at the time and costs were prohibitively expensive. The situation is very different now, however, as relatively inexpensive, off-the-shelf AI tools are now widely available and a fair amount of experience has been gained in the development of expert systems. Consequently, the PPP has most recently embarked on a revitalized effort to develop XPS-E using artificial intelligence techniques."[2]

The above statement shows that expert process planning system development is not only of interest to academic researchers but it also has been put into the agenda of industrial research organizations. In order to take a leading position, everyone who is involved in process planning system development should take a very sincere attitude toward the application of expert system technology in process planning.

1.7 Organization of the Book

This book is divided into six chapters. This chapter provides a general background of the subject. Chapter 2 presents the design representation methods. We know that the input to a process planning system is the design. The way a part design is represented is critical to the way a process planning system is constructed. In Chapter 3, CAD interface issues are addressed. In order to automate the process planning task, it is desirable to eliminate the manual data input to the system. Since modern industry designs are available in the CAD format, it is logical that an interface is developed to take the advantage of the availability of the data. The reader will find that although an interactive CAD interface is not too difficult to build, to automatically interpret an engineering design is not a trivial task. Much research is still needed in this area.

Chapter 4 discusses process planning knowledge. Process planning knowledge is the key to an expert process planning system. In a group technology-based process planning system, process knowledge is implicitly contained in the existing process plans. To develop an expert process planning system, such knowledge must be extracted from the process planners and represented in an appropriate form for the system to use. In the chapter, not only is general discussion given but also some useful data is provided. This data can be used to build an experimental process planning system. The following chapter, Chapter 5, focuses on how to put together an expert process planning system. Basic expert process planning system architecture is

[2] CAM-I Porcess Planning Program, *Statement of Work, Conceptual Model of Automated Process Planning for the Machined Parts Domain*, December 2, 1988.

presented. Several other issues such as process plan optimization, parameter optimization, etc., are discussed in the chapter as well.

Finally, the last chapter, Chapter 6, presents an example system—Quick Turnaround Cell (QTC). The QTC system has been built using the principles discussed in the book. System architecture, individual functional modules, algorithms used, implementation considerations, and an example run are presented in the chapter. It is used also as a conclusion to the book. References are provided at the end of each chapter. A comprehensive bibliography on process planning is given at the end of the book.

References

Ahlgrim, S.C., and Chang, T.C., "A Survey on the Usage and Development of Computer Aided Process Planning Systems," Report to the Engineering Research Center on Intelligent Manufacturing Systems, Purdue University, West Lafayette, Ind. 47907, February 1989.

Allen, K., "Generative Process Planning System Using DCLASS Information System," Monograph 4, Computer Aided Manufacturing Laboratory, Brigham Young University, Utah, 1979.

Alting, L. and Zhang, H., "Computer Aided Process Planning: The state-of-the-art survey," *Internal Journal of Production Research*, vol. 27, no. 4, pp. 553–585, 1989.

Barkocy, B.E. and Zdeblick, W.J., "A Knowledge Based System for Machining Operations Planning," SME Technical Paper MS 84-716, 1984.

Berra, P.B. and Barash, M.M., "Investigation of Automated Process Planning and Optimization of Metal Working Processes," Report 14, Purdue Laboratory for Applied Industrial Control, West Lafayette, Ind., July 1968.

Berra, P.B., and Barash, M.M., "Investigation of Automated Process Planning and Optimization of Metal Working Processes," Report 14, Purdue Laboratory for Applied Industrial Control, West Lafayette, Ind., July 1968.

Brown, P.F. and Ray, A., "Research Issues in Process Planning at the National Bureau of Standards," in Proceedings of the 19th CIRP International Seminar on Manufacturing Systems, Pennsylvania State University, 1987, pp. 111–119.

Chang, T.C. and Wysk, R.A., *An Introduction to Automated Process Planning Systems*. Prentice Hall, Englewood, N.J., 1985.

Chang, T.C., "TIPPS: A Totally Integrated Process Planning System," Ph.D. Thesis, Virginia Polytechnic Institute and State University, Blacksburg, Va., 1982.

Choi, B.K., "CAD/CAM Compatible Tool Oriented Process Planning for Machining Centers," Ph.D. Thesis, Purdue University, West Lafayette, Ind., 1982.

Cincinnati Milacron "Intelligent Machining Workstation Initiative," report, January 1989.

Cutkosky, M.R., Tenenbaum, J.M., and Muller, D., "Features in Process-Based Design," Proceedings of the 1988 ASME International Computers in Engineering Conference and Exhibition, San Francisco, Calif., July 1988. pp. 1–13.

Descotte, Y. and Latombe, J.C., "GARI: A Problem Solver That Plans to Machine Mechanical Parts," in Proceedings of IJCAI-7, 1981, pp. 766–772.

Englert, P.J. and Wright, P.K., "Principles for part setup and workholding in automated manufacturing," *Journal of Manufacturing Systems*, vol. 7, no. 2, 1988, pp. 147–161.

Eversheim, W. and Esch, H., "Automated generation of process plans or prismatic parts," *Annals of CIRP*, vol. 32/1/83, 1983, pp. 361–364.

Eversheim, W., Holz, B., and Zons, K.H., "Application of Automatic Process Planning and NC Programming," in Proceedings of AUTOFACT WEST, Society of Manufacturing Engineers, Anaheim, Calif., 1980, pp. 779–800.

Giusti, F., Santochi, M., and Dini, G. "COATS: An expert system for optimal tool selection," *Annals of CIRP*, vol. 35/1/86, 1986, pp. 337–340.

Haas, M. and Chang, T.C., "A Survey on the Usage of Computer Aided Process Planning Systems in Industry," Engineering Research Center on Intelligent Manufacturing Systems, Purdue University, W. Lafayette, Ind., 1987.

Henderson, M.R., "Extraction of Feature Information From Three Dimensional CAD Data," Ph.D. Thesis, Purdue University, West Lafayette, Ind., 1984.

Hummel, K.E. and Brooks, S.L., "Symbolic Representation of Manufacturing Features for an Automated Process Planning System," in Bound Volume of the Symposium on Knowledge Based Expert Systems for Manufacturing, pp. 233–243, The Winter Annual Meeting of the ASME, Anaheim, Calif., Dec. 7–12, 1986.

Iwata, K., Kakino, Y., Oba, F., and Sugimura, N., "Development of Non-Part Family Type Computer Aided Production Planning System CIMS/PRO," in Advanced Manufacturing Technology, ed. P. Blake, pp.171-184, North-Holland Publishing Co., Amsterdam, 1980.

Jakubowski, R., "Syntactic characterization of machine parts shapes," *Cybernetics and Systems: An International Journal*, vol. 13, 1982, pp. 1–24.

Joshi, S., and Chang, T.C., "Graph based heuristics for recognition of machined features from a 3-D solid model," *Computer Aided Design*, vol. 20, no. 2, 1988, pp. 58–66.

Joshi, S., Vissa, N.N., and Chang, T.C., "Expert process planning system with solid model interface," *International Journal of Production Research*, vol. 26, no. 5, 1988, pp. 863–885.

Kakino, Y., Ohba, F., Moriwaki, T., and Iwata, K., "A New Method of Parts Description for Computer Aided Process Planning," in Advances in Computer-Aided Manufacturing, ed. D. McPherson, pp. 197–213, North-Holland Publishing Co., Amsterdam, 1977.

Kramer, T.R., and Jun, J.S., "Software for an Automated Machining Workstation," NBS report, July 1986

Kumara, S.R.T., Joshi, S., Kashyap, R.L., Moodie, C.L., and Chang, T.C., "Expert systems in industrial engineering," *International Journal of Production Research*, vol. 24, no. 5, 1986, pp. 1107–1125.

LaTombe, J.C., and Dunn, M.S., "XPS-E: An expert system for process planning," Proceeding of CAM-I's 13th Annual Meeting and Technical Conference, Clearwater Beach, Fla., Nov. 13–15, 1984.

LaTombe, J.C., Tsang, J.P., Sack, C.F., Weisser, P.T., and Dunn, M.S., "XPS-E Phase 2A Final Report," CAM-I publication R-85-PPP-02, July, 1985.

Li, J., Han, C., and Ham, I., "CORE-CAPP A Company-Oriented Semi-Generative Computer Automated Process Planning System," in Proceedings of the 19th CIRP International Seminar on Manufacturing Systems, Pennsylvania State University, June 1987, pp. 219–225.

Matsushima, K., Okada, N., and Sata, T., "The integration of CAD and CAM by application of artificial intelligence techniques," *Annals of CIRP*, vol. 31/1/82, 1982.

Mouleeswaran, C.B., "PROPLAN: A Knowledge Based Expert System for Manufacturing Process Planning," Master's Thesis, University of Illinois at Chicago, 1984.

Nau, D.S. and Chang, T.C., "A Knowledege Based Approach to Process Planning," in Bound Volume of the Symposium on Computer Aided/Intelligent Process Planning, pp. 65-72, The Winter Annual Meeting of the ASME, Miami Beach, Fla., Nov. 17–2, 1985.

Nau, D.S. and Chang, T.C., "Prospects for process selection using AI", *Computers in Industry*, vol. 4, no. 3, 1983, pp. 253–263.

Nau, D.S. and Gray, M., "SIPS: An Approach of Hierarchical Knowledge Clustering to Process Planning," The ASME Winter Annual Meeting, Anaheim, Calif., Dec. 1986.

Niebel, B.W., "Mechanized Process Selection for Planning New Designs," ASTME Paper 737, 1965.

Nolen, J., *Computer Aided Process Planning for World-Class Manufacturing*, Marcel Dekker, New York, 1989.

OIR, MULTIPLAN, Organization for Industrial Research, Inc., Waltham, Mass., 1983.

Opitz, H., *A Classification System to Describe Work Pieces*, Pergamon Press, Elmsford, N.Y., 1970.

Phillips, R.H., "A Computerized Process Planning System Based on Component Classification and Coding," Ph.D. Thesis, Purdue University, West Lafayette, Ind., 1978.

Pinson, L.J. and Wiener, R., *An Introduction to Object-Oriented Programming and Smalltalk*, Addison-Wesley, Reading, Mass., 1988.

Preiss, K. and Kaplansky, E., "Automated CNC Milling by Artificial Intelligence Methods," in Proceedings of AUTOFACT 6, 1984, pp. 2.40–2.59.

Scheck, D.E., "Feasibility of Automated Process Planning," Ph.D. Thesis, Purdue University, West Lafayette, Ind., 1966.

Tulkoff, J., "Lockheed's GENPLAN," Proceedings of 18th Numerical Control Society Annual Meeting and Technical Conference, Dallas, Texas, 1981, pp. 417–421.

Wang, H-P. and Wysk, R.A., "An expert system for machining data selection," Computers and Industrial Engineering, vol. 10, no. 2, 1986, pp. 99–107.

Wysk, R.A., "An Automated Process Planning and Selection Program: APPAS," Ph.D. Thesis, Purdue University, West Lafayette, Ind., 1977.

2

Design Representation

In this chapter product design representation methods are discussed. The discussion focuses especially on the useful representations for automated process planning purposes. As all of us are aware, manufacturing is a means to realize the design. Design is a process which expresses a design requirement by a physical entity. This physical entity must functionally satisfy the design requirement. At the design stage, the result of the design process is a concept expressed in a communicable media. Often we call this concept, which is embedded in a communicable media, the design. In order not to get confused with other uses of the term design, in this chapter we will use design representation to denote this design in a communicable media.

In ancient times, a design representation was done using free hand sketches. One can find quite a few ancient designs represented in this form. From Leonardo da Vinci's sketch pads we have found a large number of very ingenious designs. Most of them are sketches of objects in perspective projection. Given da Vinci's artistic talent, those design sketches are not only engineering designs but also valuable art. In a sketch, in order to make clear some details, lengthy texts are often added. As long as the idea can be passed along to others completely and correctly, the representation is acceptable. This form of design representation was used for centuries. Even today, when we open popular magazines we still see lots of simple designs represented in sketches.

Unfortunately, not everyone has such artistic talent—we aren't always able to draw a clear and precise sketch of our idea. Also, such sketches work fine for relatively simple objects; however, when we start dealing with machines of today's sophistication, the sketch approach is no longer adequate. Obviously a new tool is needed. Since the industrial revolution, a new way of design representation has been emerging. The new method is engineering drafting. A set of universally agreed drafting symbols and rules has been developed. Multiple view projections have been adopted as the representation method. The geometry of the design is projected onto as many views as needed to obtain clarity. Cut-off views are also allowed to add more details to the representation. For very complex surfaces, section views are used to show slices of the surface. Standard annotation, which includes dimensions, tolerances, etc., is used. Now the design representation is no longer an art, but a science. Drafting is a scientific way of representing an engineering object. Projective geometry is the science underlining it. Engineers and crafters are trained to read and to prepare engineering drawings. They need to be able to reconstruct the complete design in their mind from a drawing. There is a coding (into drafting) and a decoding (interpreting) process associated with the preparing and reading of a drawing. For a complex object, several drawings may be used to show the overall shape and details.

Since paper is two dimensional, when a design is represented on a sheet of paper, projections, views, and sections are needed to represent the overall shape and the details. Since the invention of computer graphics in the 1950s, we have no longer been bounded by the dimensionality of the paper. However, initially, the computer was used as an electronic drafting board. The exact same drawing was encoded electronically. What is represented by the computer is a metaphor of the pencil and ink drawing.

On the manufacturing side, geometries of the design are used for preparing fixtures and tools. The annotation is important for material selection, processes selection, inspection, and special handling of the workpiece. With the design in electronic format, the geometries can be used for NC cutter path generation. By specifying cutter motion, a computer-assisted part programming system can use the geometries to calculate cutter locations. A computer-aided drafting system is therefore adequate for two dimensional and two and one half dimensional machining. When three dimensional machining is needed, a 3D wireframe CAD system, then, is necessary. The majority of CAD systems used today are either 2D drafting or 3D wireframe systems.

In real life, many engineering objects have free form surfaces (sculptured surfaces). The body of an automobile, the exterior of a telephone, the hull of a ship, and the fuselage of an airplane—one can go on and on—they all have curved surfaces. Traditionally, these curved surfaces are represented by sectional curves. The processes of constructing the entire surface from these sections is called lofting. Templates of these sectional curves are used to compare with the finished surface in order to determine how much more material needs to be removed from the workpiece. It is a very tedious process to go through. Surface modeling techniques such as Coons' patch [Coons 1967], Ferguson's surface [Ferguson 1964], Bezier's surface [Bezier 1972], B-spline surface, and recently NURB (Non-Uniform Rational B-spline) surface [Ding & Davies 1987] were developed to model those free form surfaces.

In order to remove hidden lines from a 3D wireframe model, face information is needed. After adding face information, CAD systems were able to generate realistic pictures of objects. This improvement helped the rendering of pictures, yet, did not contribute much to the manufacturing applications. In manufacturing, what lies beyond interactive NC part programming are automatic NC part program generative, machining process planning, and assembly planning. In order to achieve the latter, a complete and unambiguous model of the product is needed.

The development of 3D solid modeling began in the earlier 1970s. Some of the early systems are PADL [Voelcker and Requicha 1977], BUILD-1 [Braid 1973], TIPS-1 [Okino 1973], GLIDE [Eastman 19], etc. More history of solid modeling can be found in Baer et al. [1979] and Requicha and Voelcker [1982]. Some systems such as Para Solid (formerly Romulus (based on BUILD), UniSolid (based on PADL-2), GMSolid (also based on PADL-2), etc., are available commercially. Due to its completeness and uniqueness of representing 3D object, solid modeling has been considered as the most useful CAD tool for the future. The requirements of a large computing resource, the modeling difficulty, and the lack of a unified tolerance model are some of the problems which have delayed the full introduction of solid modeling in the workplace.

To ease the modeling difficulty, in recent years a feature based modeling concept has been proposed. Using features that are application oriented to model parts or products is much easier than conventional modeling methods. A feature carries not only geometry and topology, but also the semantics of the entity. This property makes the application design easier. A feature-based design system is usually a front end to a solid modeler. The geometry and the topology of a design are maintained by the solid modeler. An additional feature model is kept in the system and can be evaluated into a solid model when needed.

In this chapter, we will discuss different ways a part can be represented. We will emphasize the representations which are commonly received by a process planner. As well, the pros and cons of these representations from a process planning point of view will be discussed. Since the objective of this book is to discuss the expert process planning system, the design representations which can be used directly by an expert process planning system are discussed in more detail. In Section 2.2.1, the basic part representation methods are discussed. Sections 2.2.2–2.2.4 discuss representations which are better for expert process planning systems. The interface issues are discussed in Chapter 3.

2.1 Basic Part Representation Methods

As discussed previously, there are many ways a product can be represented. Not every representation is appropriate, however, for manufacturing applications. In this section, we would like to show commonly used methods of representing a product. All the methods discussed are currently being used. For automated process planning purposes, one needs to select the presentation that is complete in the modeling domain and is also easy to use for planning.

2.1.1 Natural Language Description

One method that is most inaccurate, yet is at times easiest to use is natural language description. Natural language means a language which is spoken daily by humans. It can be the English language of this book or any foreign language. Natural language is generally not very precise when compared to artificial languages such as computer languages. However, everyone, engineer or nonengineer, understands it. For very simple parts, quite often we don't even bother to prepare a drawing. We can simply describe it to other people using words. Of course we must be sure that the recipients of our description will understand it as we do. Many novelists can give a vivid description of machines of the future. Movie makers can even produce devices based on the descriptions. However, not all of us have the talent for painting a detailed picture of machines using words. Therefore, the engineering object described by us using words should be simple. One example is a description of a part "rod":

> The rod should be made of low carbon steel. The finished rod has 1″ diameter and 10″ length. Both ends should be rounded. At one end, approximately 1″ from the end, drill a 3/8″ hole.

The above description is almost a complete description of a part, although there are still ambiguities about the end rounding radius as well as the surface finish and tolerances of the part. In general, we have to assume that these missing parameters are not critical. Thus, no attention should be placed on them. With the above description, a machinist should be able to make the part.

To describe a simple rod is easy. But how about describing a cam using natural language. Very quickly we will find that natural language is not sufficient for engineering design. It is thus obvious that natural language description is not a candidate for automated process planning.

2.1.2 Freehand Sketch

A freehand sketch is actually the most common method used by designers. In most design departments, designers prepare only sketches. The sketches are then given to the drafters. Drafters prepare the drawing. A freehand sketch is used most frequently during the conceptual design stage. A 3D projection is the most frequently used method of preparing the sketch. The main goal of the sketch is to pass the design to the drafter for preparing a formal drawing. Therefore, the essential information, such as dimension, tolerance, surface finish, etc., must be present in the sketch (Fig. 2.1). There must be a good understanding between the designer and the drafter. When ambiguity arises due to incomplete specification in the sketch, the drafter has to consult the designer for clarification of the design. Quite often, drafters are given the responsibility to decide the details. For very simple and unimportant engineering parts that are used only once, the sketch may be prepared and passed to the machinist for making the part. Otherwise, a formal drawing is prepared from the sketch.

One advantage of using a sketch is the ease of preparing it. Drawing rules do not need to be strictly followed. The designer is relieved from the tedious and time-consuming drafting or modeling work. More effort can be spent by the designer on the creative part of the design work. However, the major disadvantage of sketches is

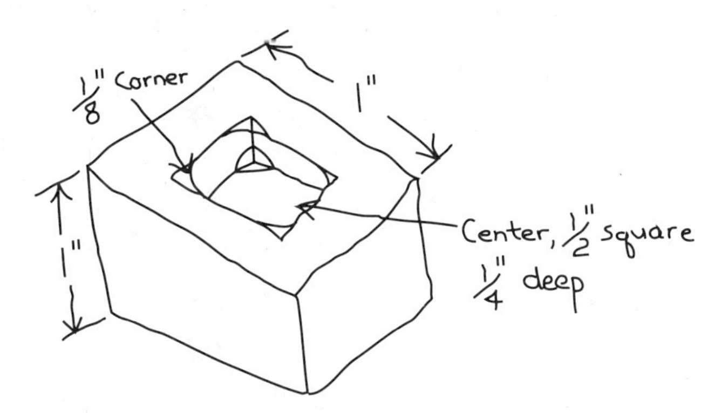

Figure 2.1 A free hand sketch

the lack of precision caused by not following strict drawing rules. Since designers are not always artistic, their sketches may not be understood by everyone. A freehand sketch should not be issued as the production drawing. The sketch is not very useful for automated process planning either. It needs to be translated by a human planner into an electronic format before it can be used.

2.1.3 Engineering Drafting

Engineering drafting is a formal way of representing an engineering object. A set of standard rules are applied to prepare the drawing. These rules are viewed as a universal language for engineers. The rules include the convention of geometry, dimensioning, and tolerancing. Additional annotations may be placed; however, there is no standard way of formatting them. The standards are company specific. The geometry of drafting consists of lines, circles, and curves. Visible edges of the part are shown in solid lines, and invisible edges are shown in dashed lines. Sometimes when the part is too complex, however, the dashed lines are not shown. To help further clarify the drawing, center lines, hatching lines, etc., are also used. Center lines define the center of a symmetric geometry. Hatching lines are used to show cross sections. Different hatching lines denote different types of workpiece materials used.

Since drafting uses two dimensional drawings to represent three dimensional objects, projections are used. The standard of engineering drafting is to use multiview (front, top, side, sections) orthographic projections to represent an engineei.ng object (Fig. 2.2). For a typical three view drawing, the three views are orthographic projections of the object on XY, XZ, and YZ planes. If one considers the space divided by the XY and XZ planes at the +X half space (+X half space is the space where $x > 0$), there are four quadrants (angles). The +Z+Y quadrant is called first angle. The −Z−Y quadrant is called third angle. In the United States, a third angle projection is the standard. Third angle projection means the object is placed in the third angle. The view on XY plane is the front view, on the XZ the top view, and on the YZ the side view.

The section views can be taken on any auxiliary projection plane. Again, the projection is always orthographic. Sections are used for clarifying the details of the part. The auxiliary projection planes are shown on the view where it is perpendicular to the view. It is shown as a line segment with arrows at the end of the line segment denoting the view direction. Section designator, such as C-C, is marked on the arrows. A drawing, either on the same sheet of paper or on different paper, shows the section projection. There can be as many sections as necessary to show the details.

The number of views used is the minimum which can still pro-

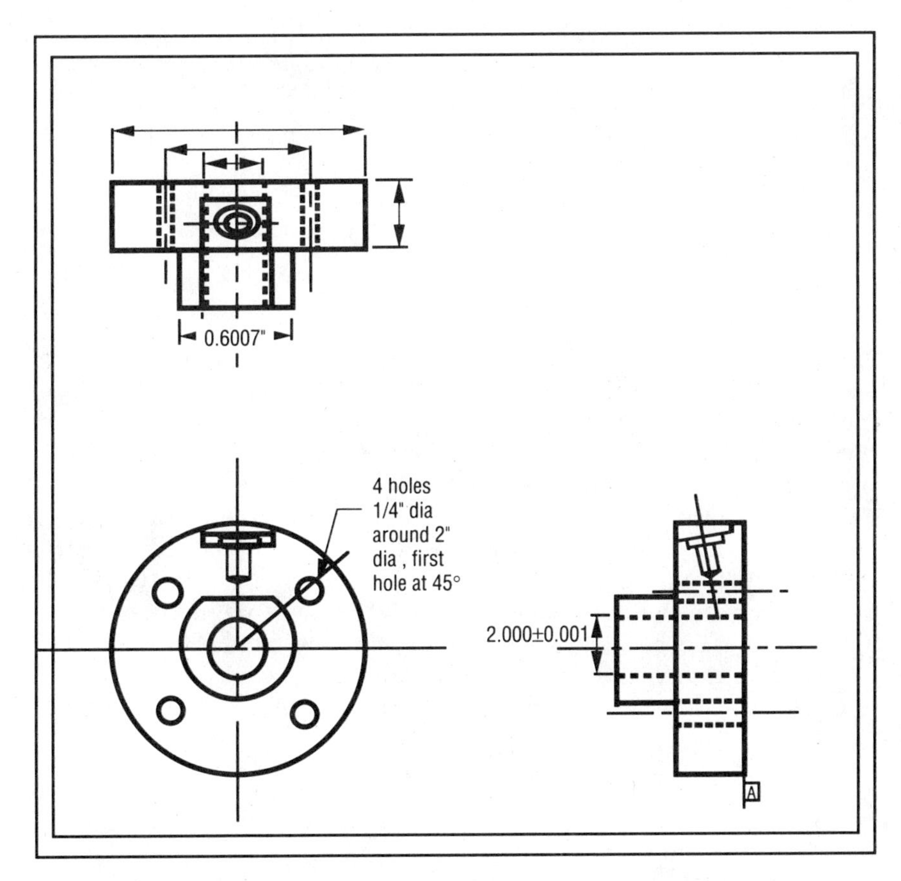

Figure 2. 2 An engineering drafting

vide complete information of the part. For example, a very simple turned part can be represented by just one partial view (Fig. 2.3). In the figure, it can be seen that the part has a flange and a center bore. The center line (dashed line) indicates that the part is symmetric around the center line (a rotational part). The hatching lines denote the solid material section, the area above the hatching area is a hole. With this simple drawing, the complete geometry of the part can be interpreted. For such a simple part, one view is sufficient to show the complete geometry. However, for most engineering parts, a three view drawing is the standard.

To interpret a multi-view drawing, one starts from one view—say, the front view. The front view is swept according to the top view to make an initial solid. For example, in Fig. 2.4 the line *a* in the top view represents the front view itself. Each of the lines *b, c, d,* and *e* swept the front view to make the solid in Fig. 2.5(a). The slide lines in the front view are swept into solid faces; the dashed line is swept

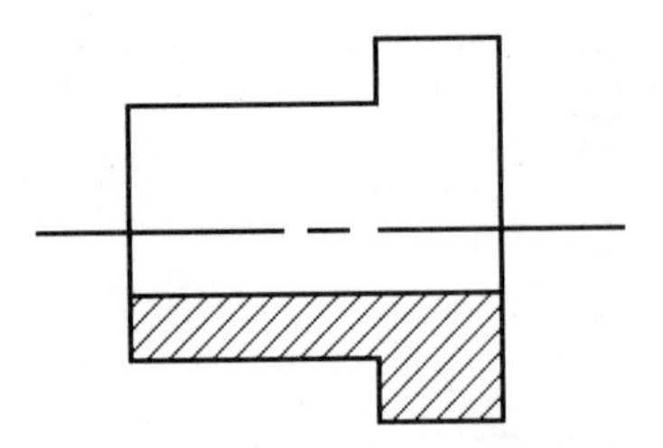

Figure 2.3 Partial view of a simple turned part

into a dashed face. A dashed face is not a complete face, part of it should be deleted. We also need to keep lines *f* and *g* on the top of the solid, because they represent either entrance faces to a cavity or the side wall of a protrusion. Whether the feature is a cavity or a protrusion can be found in either the front view or the side view. The dashed face in Fig. 2.5(a) already shows that the feature is a cavity. Now the *h* and *i* lines in the side view can be used to resolve the location of the feature. They sweep lines *f* and *g* into two solid faces as shown in Fig. 2.5(b). The lines *d*, *e* in between these faces are deleted. The dashed face is cut by the new faces. Since it is the bottom of a cavity, the dashed faces outside the cavity are deleted, and the one in between is turned into a solid. A slot is formed (Fig.

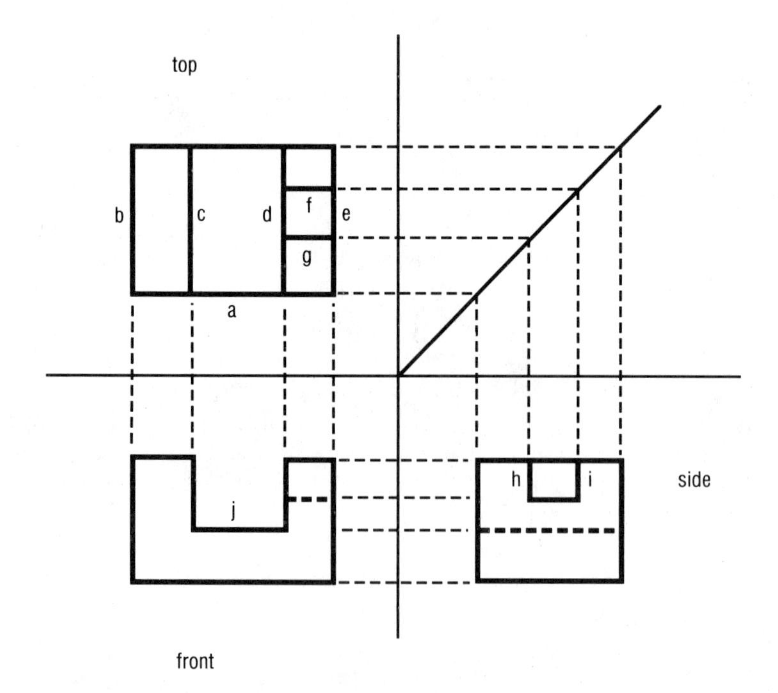

Figure 2.4 Projections of a part

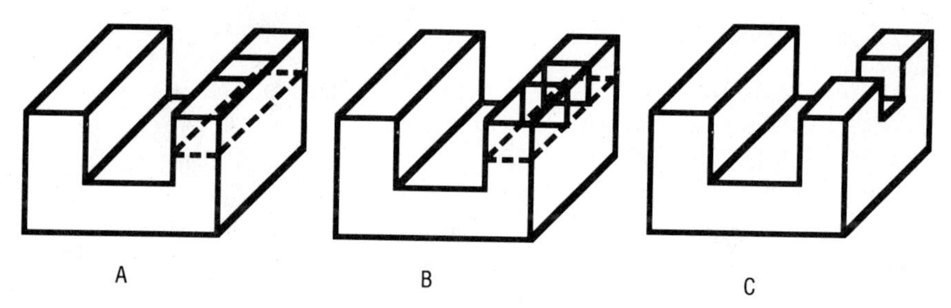

Figure 2.5 Drawing interpretation

2.5(c). Finally, the dashed line in the side view is checked against the front view. Line *j* agrees with the dashed line in the side view. Since there is no conflict, the final part geometry is the one represented in the drawing. The entire interpretation process is conducted in one's head. Projective geometry is the foundation of the drawing interpretation.

Drawing interpretation is not difficult for engineers to do; however, to do it automatically poses many problems. In addition to the basic geometry lines, there are many other lines, such as dimension lines, a center line, pointers for the notes, etc. It may not be easy to eliminate them correctly. A complex drawing may have several thousand lines and curves that are not always drawn to precise scale. Sweeping may not create a legible final geometry. To resolve the hundreds of conflicts may not be a tractable problem. To recognize manufacturing features from the drawing automatically is even more difficult. Although engineering drafting is still a standard method of representing engineering objects, it cannot be used directly by a process planning system.

2.1.4 Physical Models—clay model, template

For many products, such as automobile bodies and cases of home appliances, etc., which have sculptured (free form) surfaces, clay models are commonly used to represent the design. This physical model is either full scale or a smaller scale. The look of the shape is more important than the functionality of the surface. For manufacturing, the clay model is digitized and points are used to fit a certain curve. This curve is then either used to prepare a sectional drawing of the part or used directly for an NC part program.

For inspection or non-NC manufacturing, the sectional drawings are made into templates for comparison purposes. The workpiece is compared with the template to ensure the designed shape is inscribed in the workpiece correctly. The physical model is an intermediate design representation media. It cannot be directly used in an automated process planning system.

2.1.5 Surface Model

A modern way of representing sculptured surfaces is to use computer models. As discussed in the introduction, there are several sculptured surface models. Basically, a surface is represented by surface patches. A surface patch (Fig. 2.6) is a small surface area. Each surface patch is modeled using the same mathematical model, an n-degree polynomial parametric model. A polynomial parametric model can be written as the following:

$$r(t,s) = \sum_{i=0}^{n} \sum_{j=0}^{m} r_{ij} B_{i,n}(t) B_{j,m}(s)$$

$$r(t,s) = \begin{vmatrix} x(t) \\ y(t) \\ z(t) \end{vmatrix}$$

where

r_{ij}: control points
t,s: two parameters
$B_{i,n}(t)B_{j,m}(S)$: blending functions
n, m: numbers of control points minus one on the t and s directions, respectively

A surface patch is defined by $(n + 1) \times (m + 1)$ control points. For Bezier's surface, the number of control points determines the degree of the polynomial. When there are 2×2 control points, the surface is planar. With 3×3 control points, a curved surface, either convex or concave, can be defined. Using 4×4 control points, an enclosed surface, or surface with reflecting point can be defined. The end points and end tangents can also be specified. When more than 4×4 control points are used, a higher degree (> 3) polynomial model is used that is more flexible for defining surface; however, it

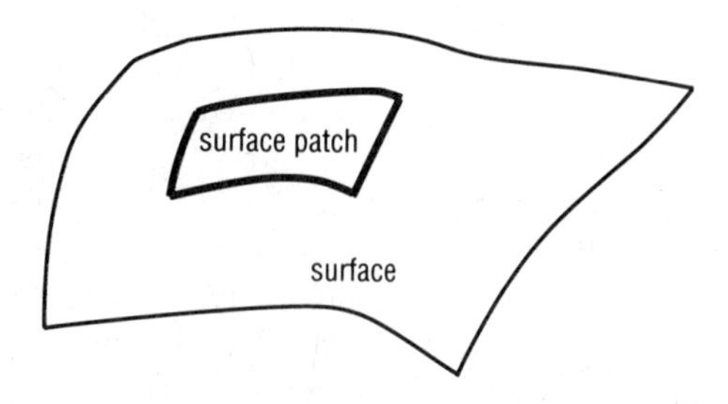

Figure 2.6 Surface patch

becomes much more difficult for computation. Therefore, for Bezier surface, generally 16 points ($n = m = 3$) are used. Such a surface is called a bi-cubic surface, since the polynomial is of third order and the surface has two parameters. The blending functions are cubic functions. For example, for Bezier surface, a curve at ($s = 0$) can be written as follows:

$$r(t) = \sum_{i=0}^{3} r_i B_{i,3}(t)$$

$$B_{i,3}(t) = C(3,i)t^i(1 - t)^{3-i}$$

$$C(3,i) = \frac{3!}{i!(3 - i)!}$$

Figure 2.7 shows the blending functions for a cubic Bezier's model. From the equation above, $r(t)$ is equal to the summation of r_i times the corresponding blending functions $B_{i,3}$. It can be seen that $B_{i,3}$ has a value greater than zero for the entire range of t ($0 \leq t \leq 1$). Therefore, any control point r_i has an influence on the overall curve. Changing the position of any control point will affect the shape of the entire curve/surface.

For the B-spline surface, the degree of the polynomial blending function is determined by a parameter k and is usually independent of the number of control points. The B-spline blending functions are defined recursively by the following equation:

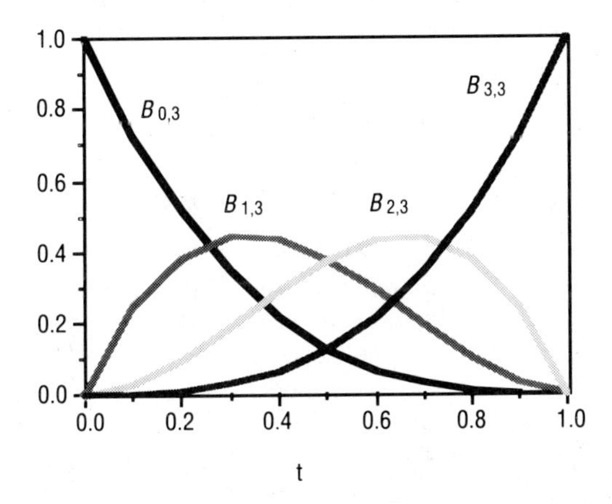

Figure 2.7 Value of Bezier's blending functions (cubic Bezier's model)

$$B_{i,1}(t) = 1 \text{ if } l_i \leq t \leq l_{i+1}$$
$$= 0 \text{ otherwise}$$

and

$$B_{i,k}(t) = \frac{(t - l_i)\, B_{i,k-1}(t)}{l_{i+k-1} - l_i} + \frac{(l_{i+k} - t)\, B_{i+1,k-1}(t)}{l_{i+k} - l_{i+1}}$$

$$\begin{aligned} l_i &= 0 & &\text{if } i < k \\ l_i &= i - k + 1 & &\text{if } k \leq i \leq n \\ l_i &= n - k + 2 & &\text{if } i > n \end{aligned}$$

with

$$0 \leq i \leq n + k$$

The range of the parameter t is

$$0 \leq t \leq n - k + 2.$$

Again, usually a three degree polynomial is used, $k = 3$. B-spline provides better local control of a surface. In the B-spline bi-cubic surface, the influence of the control points is regional (Fig. 2.8). More control points can be added without affecting the degree of the polynomial. However, in Bezier's surface, changing one control point affects the shape of the entire surface. The closer to the changed control point, the greater the change to the surface. Because of this property, B-spline is considered superior than Bezier's surface.

It is important to point out that the control points do not lie on the surface. Rather, they form a characteristic polygon (Fig. 2.9) which covers the surface defined. A design made with a sculptured surface processor usually begins with the designer selecting control points interactively on a CRT terminal. According to the surface geometry plotted by the computer, the designer modifies the surface by moving the control points. Both Bezier's surface and the B-spline surface are suitable for interactive design. More details can be found in reference Faux and Pratt [1979] and Mortenson [1985].

Since sculptured surfaces are defined explicitly and the only process to machine a sculptured surface is ball-end milling, there is not much problem for process selection. However, to determine the roughing and finishing of the surface, to select the appropriate cutter size, and to do fixturing method planning are difficult problems. Currently, most process planning systems do not deal with sculptured surface planning. Sculptured surfaces are usually handled by an interactive NC part programming system.

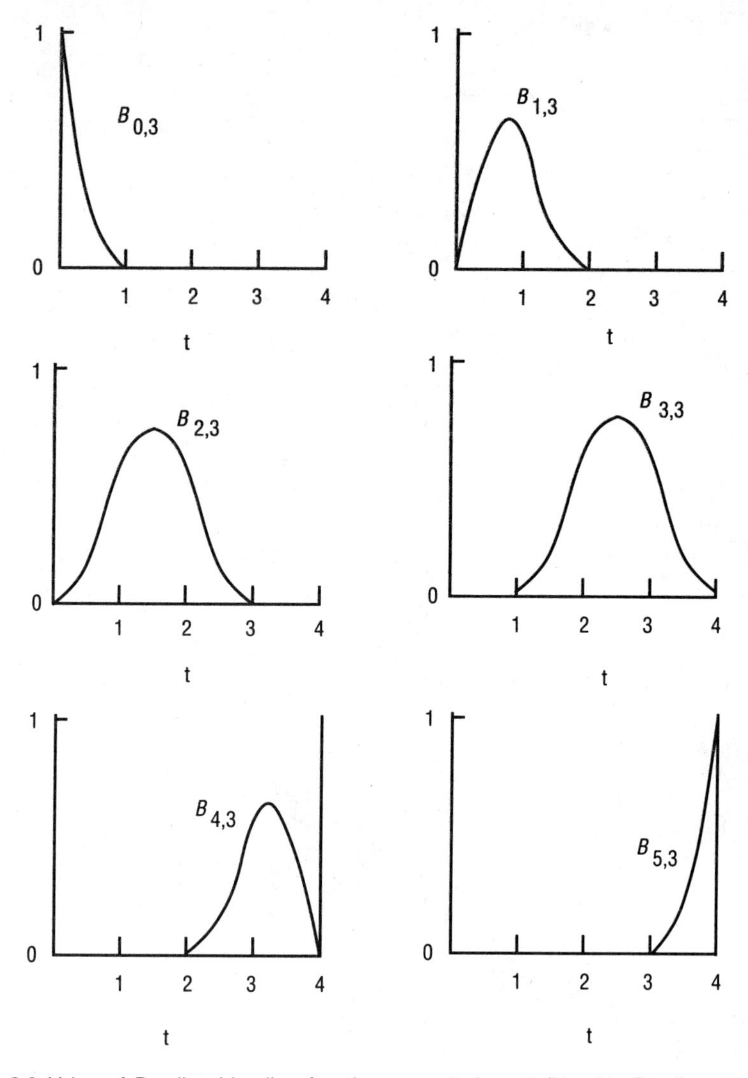

Figure 2.8 Value of B-spline blending function, $n = 5$, $k = 3$ (bi-cubic B-spline, 5 control point per curve)

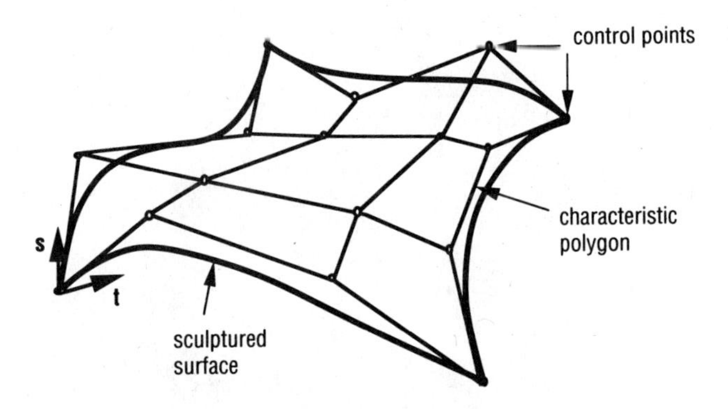

Figure 2.9 Control points and characteristic polygon

2.1.6 GT Code

All the part representation methods discussed so far dealt with detailed representation. Group Technology (GT) by its definition is a method of utilizing the similarities of parts to simplify the problem. GT coding is intentionally designed to be gross. A code represents a large number of similar parts. Generally a GT code consists of several digits. The first digit represents the overall shape. The later the digit position, the more specific feature it represents. Most GT coding schemes use a decimal code which has 10 values for each digit. Overall shape, external shape, internal shape, functional surfaces, special features (such as thread, gear teeth, hole pattern, etc.), dimension, tolerance, and raw material shape are some of the categories commonly seen. GT coding schemes cannot be used as accurate design representation methods. In Fig. 2.10, a simple part is repre-

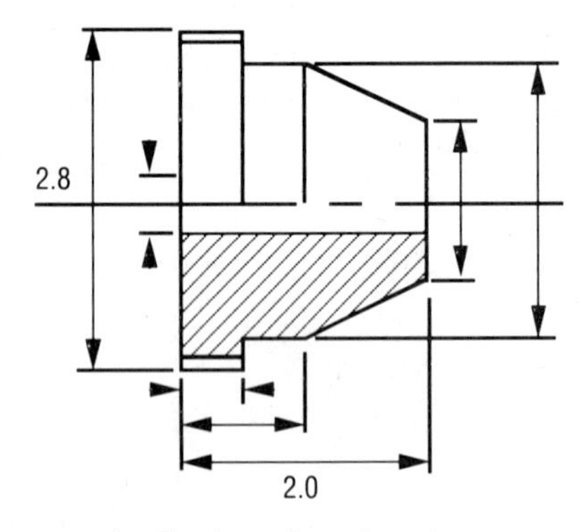

A. Engineering drawing

11106

B. GT code

Digit	Meaning	Code value
1	rotational part L/D ratio = 1.4 > 0.5	: 1
2	step to one end	: 1
3	internal bore without shape element	: 1
4	no external plane surface	: 0
5	gear teeth	: 6

C. Explanation of the Code

Figure 2.10 A part drawing and its GT code (Opitz)

sented in both drawing format and in Opitz code. The code (geometry code only) is only five digits long. It by no means provides any detailed description of the part. However, by reading the code, one can get a rough idea about the shape of the part and some special features (gear teeth in the example) it has.

GT codes are commonly used in variant process planning systems for standard process plan retrieval [Chang and Wysk 1985]. Parts are coded and stored in a database. Since GT code represents the gross shape of the part, it is possible to retrieve all the similar parts previously produced. Those similar parts are said to belong to a part family. A standard process plan can be prepared by summarizing the process plans of the part family members. Although there are a few research studies on generating GT code automatically from a solid model [Kaprianou 1980, Henderson, 1986], usually, a human planner needs to code the part before using a computer to retrieve the standard plan. Since GT code is not detailed, a formal drawing must be available to the process planner for plan modification. There are also systems [Wang and Wysk 1985] which utilize GT code to retrieve a family process model. The process model, in turn, can be used to generate process plans for the family members. In this case, the system can be considered to be generative. However, the data representations used are of two different kinds, the first one is a GT code, and the second one is a user dialogue.

GT code can also be used in design retrieval. All previous design drawings or models can be indexed using a GT code and its part number. Before a new design is composed, the design concept is coded using a GT coding scheme. This code is then used for retrieving the existing designs. The designer, then, can select one or several of the existing designs as examples for making the new design. Again, this does not automate the design process, but it helps the designer to prepare new designs. It can reduce the waste of designing from scratch.

The advantages of using a GT code to represent the part for either design retrieval or process plan retrieval are the following:

- eliminate redundant work
- standardize the design or process

However, in order to make the approach useful, the system environment must be relatively stable. Otherwise, the old design or old process plan won't be appropriate for the new environment. Another problem associated with using this approach is the hindering of human creativity. Although this can be argued, in general there is a very subtle balance between standardization and creativity. When not handled well, standardization can hinder creativity.

2.1.7 Solid Models

A solid model is a latecomer in the part representation arena. It defines the complete geometry and topology of a part. Given a model, one can mathematically determine whether a point is inside, on the boundary, or outside the object. This enables a user to compute the volume, weight, and moment of inertia of the object. These are called the volumetric properties of the model. Since the topology is in the model, one can also find the topological properties, such as connectivity and containment relationships. It is also possible through coordinate transformation to convert a solid model into traditional three view drawings.

There are several internal representations used in solid modeling. Internal representation is the way a model is represented in the computer. It is different than the external representation such as engineering drafting, sketching, etc. The external representation is what humans see on either a CRT screen or a sheet of paper. External representation is generated from an internal representation by certain transformations. Solid modeling internal representations include Constructive Solid Geometry (CSG), Boundary Representation (B-Rep), Sweeping, Spatial Occupancy Enumeration, Cell Decomposition, and Primitive Instancing [Requicha, A.A.G. 1980, Chang and Wysk 1985]. Among them the most commonly used are CSG, B-Rep, and Sweeping representations. These representations completely and unambiguously represent a solid object in the 3D space. Since the geometry and topology of the object are complete, they can be used by many engineering applications directly. They provide the most complete geometrical information for automated process planning. Therefore, several new research systems [Chang 1982, Kung 1984, Joshi, et al. 1988, Chang, et al. 1988,] use solid models as input.

Although a solid model represents an engineering object completely and unambiguously, it represents only the nominal dimension. For manufacturing, the inheritant manufacturing errors need to be recognized. Tolerance is a way to control the errors. At the time of this writing, there is still no standard way of modeling tolerances in a solid model. Only a few studies can be found. The variational geometry proposed by Requicha [1977] has been implemented [Requicha and Chen 1986] in the PADL system—a CSG modeler. CAM-I proposed to use a separate tolerance modeler coupled with a B-Rep based modeler [Johnson 1985]. The modeler is to follow the ANSI Y14.5M standard [ANSI 1973]. Turner and Anderson [1988] implemented a tolerance scheme on the feature modeler level (see Chapter 6). The B-Rep has pointers to link the faces to the tolerance specification on the feature model. Those are a few isolated cases of tolerance data implemented in a solid model. In general, tolerance information and annotations are not part of the solid modeler. Because these pieces of information are essential to manufacturing, most process planning systems require users to manually add this information to the system input. Details on CSG and B-Rep are discussed in later sections.

2.1.8 Symbolic Representation

Symbolic representation of part design is commonly used in process planning systems. Expert process planning systems take as input either a symbolic representation of the design or an interactive dialogue between the computer and a human planner. In Chang and Wysk [1985] the symbolic representation was termed special descriptive language. Several non-expert process planning systems also use symbolic representation as input. Since process planning requires feature-based description of the part design, most symbolic representations for process planning are feature-based (see Section 2.5). A feature can be a hole, a slot, a pocket, a step, etc. Most of symbolic representations cannot be used directly for drawing purposes. Therefore, they are application specific (in this case, process planning) and not general for CAD applications.

More symbolic representation methods will be introduced in Chapter 5 under the descriptive language section. Here we will show only one simple example of how to represent a part symbolically. To fully represent a part for process planning (but not necessarily for drafting), overall part description and its features need to be represented. One way to represent them is using frame representation. The top level frame can be a part frame. Each feature in the part is represented in a feature frame. Feature frames are linked to the part frame through a pointer slot. Relationships between frames are also stored in slots. An instance of a part frame may be designed as follows:

```
(
T001
(A_KIND_OF (VALUE PART))
(PART_NAME (VALUE TEST_PART))
(BATCH_QTY (VALUE 10))
(ORDER_QTY (VALUE 10))
(MTL_SIZE (VALUE (4.0 3.0 1.0)))
(MTL_TYPE (VALUE STEEL))
(MTL_SPEC (VALUE SAE_1050))
(MTL_COND (VALUE TEMPERED))
(HARDNESS (VALUE (450 500)))
```

```
(SURFACE_TREAT (VALUE NIL))
(MADE_UP_OF (VALUE (F1 F3 F4 F7 F10)))
)
```

In the part frame, information pertinent to the part is stored. The above example is implemented in Lisp language. It can be seen that there are slots for the part names, batch quantity, order quantity, material size, type, specification, condition, and hardness, and surface treatment. The last slot shows a list of top level features. Top level features are features which are not contained by other features. There are feature frames for each type of feature. For example, an instance of a feature frame for a hole can be shown as follows:

```
(
H1
(A_KIND_OF (VALUE HOLE))
(DIAMETER (VALUE .6))
(DEPTH (VALUE 1.0))
(LENGTH (VALUE 1.6))
(AXIS_DIRECTION (VALUE (0 0 1)))
(APP_DIRECTION (VALUE (0 0 -1)))
(DEPTH_REF (VALUE F18))
(ENDS_IN (VALUE F10))
(BOTTOM_TYPE (VALUE NIL))
(X_REF (VALUE F17))
(Y_REF (VALUE F16))
(X_DISTANCE (VALUE 1.0))
(Y_DISTANCE (VALUE 1.0))
(SECONDARY_FEATURE (VALUE (COUNTER BORE1)))
(CONTAINS_FEATURES (VALUE NIL))
(CONTAINED_IN (VALUE F18))
(DIAMETER_TOL (VALUE 0.001))
(DEPTH_TOL (VALUE NIL))
(SURFACE_FINISH (VALUE 32))
(POSITION_X (VALUE 0.005))
(POSITION_Y (VALUE 0.005))
(CIRCULARITY (VALUE NIL))
(CONCENTRICITY (VALUE NIL))
(CONCENTRICITY_REF (VALUE NIL))
(PERPENDICULARITY (VALUE NIL))
```

```
(PERPENDICULARITY_REF (VALUE NIL))
(HARDNESS (VALUE 500))
)
```

The advantage of this representation is the fact that it can be used directly in a process planning system.

2.2 CSG Models

As discussed in Section 2.1.7, CSG is one of the commonly used solid modeling internal representations. In a CSG model an object is represented by a binary tree consisting of geometrical primitives, transformations, and symbols representing boolean operators (Fig. 2.11). Terminal nodes can be either primitive or primitive with transformation. Nonterminal nodes can either be boolean operator or transformation. A part model is constructed using these tree building elements. CSG modelers use boolean operators to enter the model. Using boolean operators is the most versatile method in building a solid model. Even most non-CSG-based solid modelers use boolean operator.

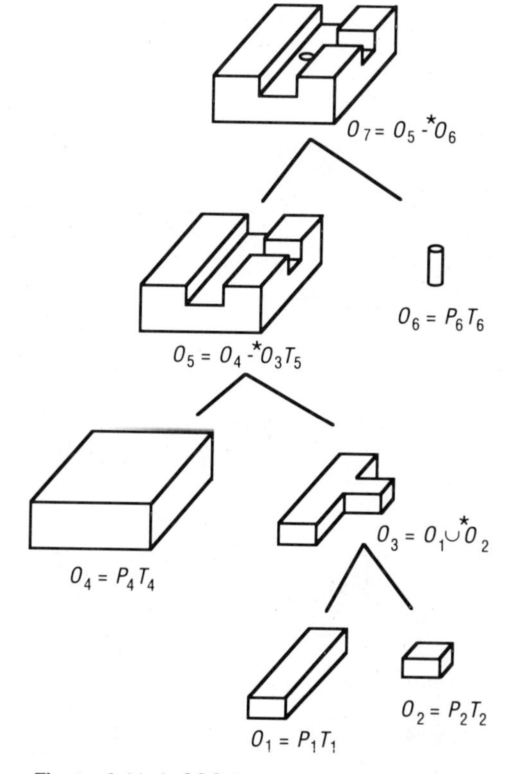

Figure 2.11 A CSG tree

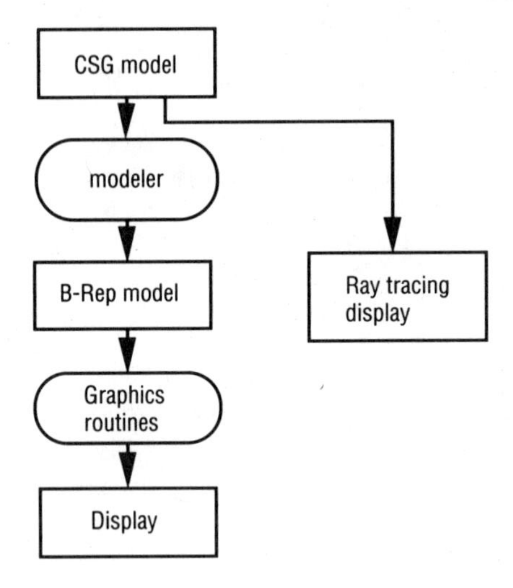

Figure 2.12 CSG operation

A CSG model can be modified by either changing the parameters of the existing tree (parameters of the primitives and transformations) or by adding new branches. The new branches can be union, difference, or intersect set operators. A sub-tree in the CSG tree represents an intermediate solid object. The validity of the solid is maintained throughout the construction of the object. Figure 2.12 illustrates the operation of a CSG modeler. The model is built either by using a text editor, a graphic editor, or a dialogue. The image can be rendered through ray tracing or by going through a boundary evaluation. The former generates an image on a raster display directly from a CSG model. The later requires several steps. First, the modeler evaluates a CSG model into a boundary representation. The boundary representation is then displayed on either a raster or a vector display by using graphics routines. The latter approach is commonly used in CSG systems.

Primitives used in a CSG modeler can be either from a fixed set of pre-defined primitives or from something created by the user (e.g., sweeping primitive in ROMULUS). Primitives can also be either solid primitives or half spaces. Commonly used solid primitives are block (parallelepipe), cylinder, sphere, torus, wedge, and cone (Fig. 2.13). Each primitive is parameterized, where the parameters define the size of the primitive. For example, the parameters for a block are length, width, and height; for a cylinder, they are radius and length. These primitives are represented internally by its bounding faces, edges, and vertices. Face types are specified by an identifier. In most CSG modelers the face types are limited to plane, cylindrical surface, spherical surface, conical surface, and torus surface. One of the

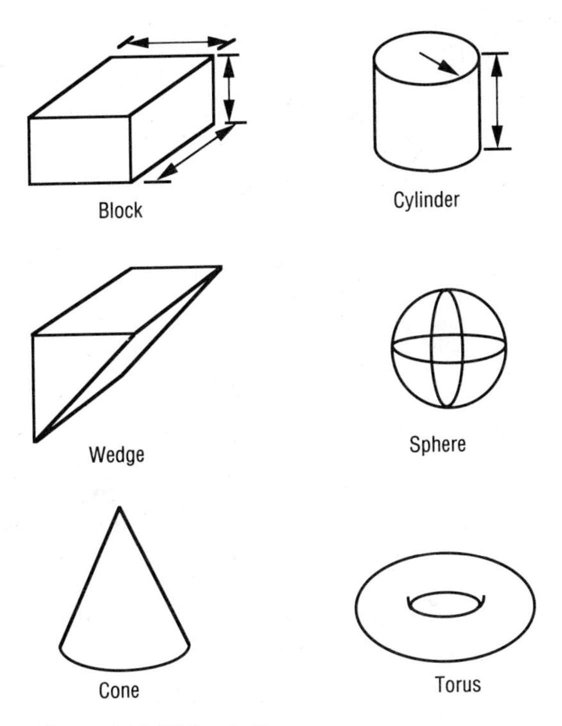

Figure 2.13 CSG primitives

difficulties of an increasing number of primitives and types of primitive surfaces is the intersection computation. The number of intersection types can be explosive. Vast numbers of computer codes will have to be written to handle all cases. Most of the systems, such as PADL-2 and ROMULUS, use a limited set of solid primitives.

Some systems (TIPS [Okino and Kubo 1973], ROMULUS also lets users define primitives using half spaces) use half space to construct a solid model. The space can be divided into two half spaces by a surface. There is the material side of the half space and the non-material side of the half space. The general form of the surface can be stated as: $f(x,y,z) = 0$. The material side of the half space is defined as: $f(x,y,z) > 0$. A convex solid can be defined by the intersection of several half spaces. The constraint is that the solid region has to be enclosed.

$$O = \bigcap_{i=1}^{n} f_i$$

where

$f_i : f_i(x,y,z) > 0.$

Figure 2.14 shows a two dimensional projection of a region

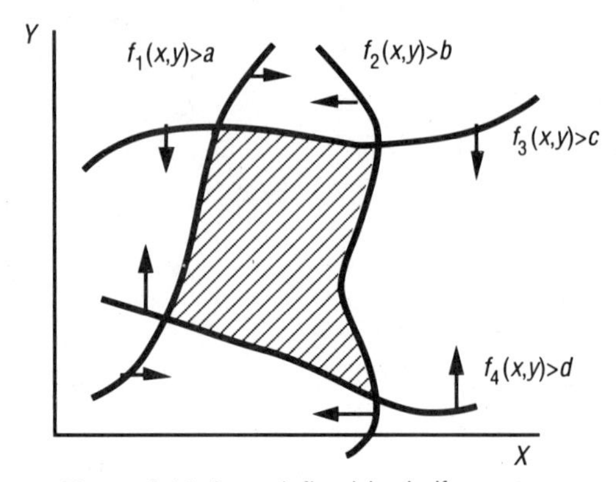

Figure 2.14 Area defined by half spaces

defined by some half-spaces. Theoretically f can be any surface function; however, there are implementation difficulties. Equations for surface intersection need to be derived and implemented in the system. It may not be possible to include every surface type.

To position and orient a primitive or an intermediate object model, a transformation matrix is used. An associated transformation matrix (T_i) defines the position and the orientation of the primitive. A complete transformation matrix is a 4 × 4 matrix (homogeneous transformation matrix). For positioning a primitive, the translation *TRAN* ($\Delta x, \Delta y, \Delta z$) is used. $\Delta x, \Delta y, \Delta z$ are the displacements of the primitive.

$$TRAN(\Delta x, \Delta y, \Delta z) = \begin{bmatrix} 1 & 0 & 0 & 0 \\ 0 & 1 & 0 & 0 \\ 0 & 0 & 1 & 0 \\ \Delta x & \Delta y & \Delta z & 1 \end{bmatrix}$$

Rotation about X, Y, and Z axes are:

$$Rot_x(\alpha) = \begin{bmatrix} 1 & 0 & 0 & 0 \\ 0 & \cos\alpha & -\sin\alpha & 0 \\ 0 & \sin\alpha & \cos\alpha & 0 \\ 0 & 0 & 0 & 1 \end{bmatrix}$$

$$Rot_y(\beta) = \begin{bmatrix} \cos\beta & 0 & \sin\beta & 0 \\ 0 & 1 & 0 & 0 \\ -\sin\beta & 0 & \cos\beta & 0 \\ 0 & 0 & 0 & 1 \end{bmatrix}$$

$$Rot_z(\gamma) = \begin{bmatrix} \cos\gamma & \sin\gamma & 0 & 0 \\ \sin\gamma & \cos\gamma & 0 & 0 \\ 0 & 0 & 1 & 0 \\ 0 & 0 & 0 & 1 \end{bmatrix}$$

Scaling can be done by:

$$SC(a,b,c) = \begin{bmatrix} a & 0 & 0 & 0 \\ 0 & b & 0 & 0 \\ 0 & 0 & c & 0 \\ 0 & 0 & 0 & 1 \end{bmatrix}$$

Rotation, translation, and scaling can all be contained in a single transformation matrix.

$$T_i = SC(a,b,c)TRAN(\Delta x,\Delta y,\Delta z)\ Rot_x(\alpha)\ Rot_y(\beta)\ Rot_z(\gamma)$$

Boolean operators include union ($\cup$), difference ($-$), and intersect ($\cap$) (Fig. 2.15). Actually, the boolean operators used in CSG construction are regularized set operators, which prevent the creation of unwanted dangling (Fig. 2.16) or disconnected parts. The differences between the ordinary boolean operators and the regularized boolean operators are well discussed in Requicha [1977] and Chapter 9 of Mortenson [1985]. Using set operators, the object in Fig. 2.11 can be written as follows:

$$C_7 = F_4T_4 - {}^*(F_1T_1 \cup {}^*F_2T_2)T_5 - {}^*F_6T_6$$

The same object can be modeled as follows:

$$C_7 - F_4T_4 - {}^*(F_1T_1T_5 \cup {}^*F_6T_6) - {}^*F_1T_1T_5$$

One can even select a smaller base block, then union it with the three small blocks on top of the part (Fig. 2.17). The resultant object is then different from a cylinder. Although the object they represent is identical, the new model is very different than the ones presented above. For the same part, there are an infinite number of ways a CSG model can be built. The above model when written in a PADL like language is the following:

```
C1 = BLOCK(1,3,.5) AT ( 1,0,.5)
C2 = BLOCK(1,1,.5) AT (2,1,.5)
C3 = C1 .UN. C2
```

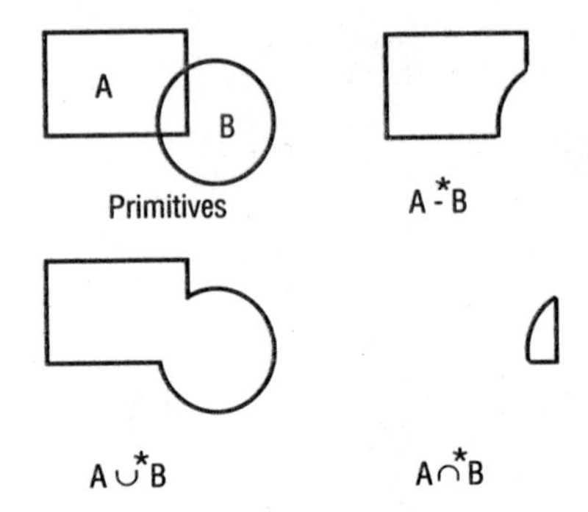

Figure 2.15 Set operators

C4 = BLOCK(3,3,1) AT (0,0,0)

C6 = CYLINDER(.25,1) AT (1.5,1.5,.5)

C5 = C6 .DIF. C3

C7 = C5 .DIF. C6

The binary tree representation is the logic representation of CSG. The above representation is the input format. The internal data structure may be something different. However, they all carry the same information. What a user sees is the input format and the display. As mentioned before, to modify a CSG model, one has to go back to the input data. Since face, edge, and vertices information are not represented explicitly, interactive editing of local geometry is not possible. When using a CSG model for an application directly, one must work with the primitives, transformations, and operators. No explicit information on geometry and topology is available in the model.

A sequence of machining operations can be modeled by a CSG

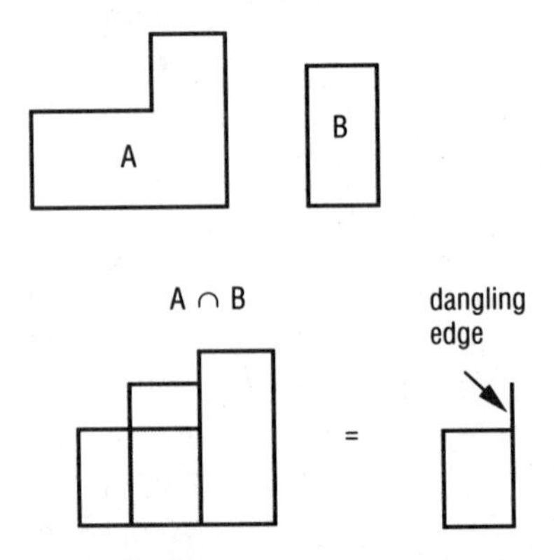

Figure 2.16 Dangling edge created by a intersect operation

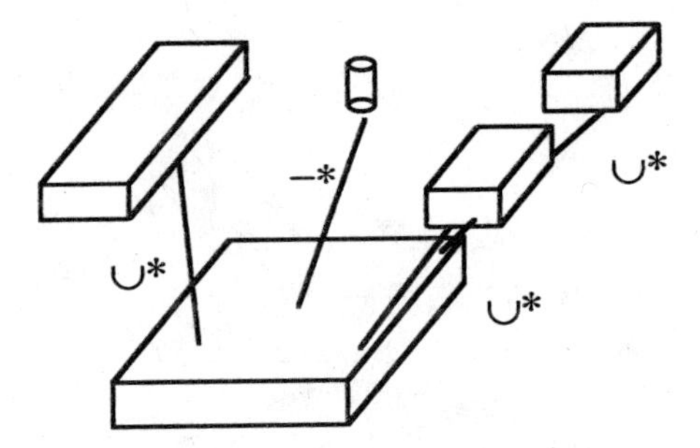

Figure 2.17 Alternative CSG model for part in Figure 2.11

model. The raw material is a base part C_1. Each tool motion can be modeled by the tool sweep volume V_i. The tool sweep volume is best modeled by using a sweep operator, $SW\ (f_i(x,y,z))$, which is available in some CSG modelers. $f_i(x,y,z)$ defines a sweep path. $V_i = C_i SW\ (f_i(x,y,z))$ C_i is the CSG model of the tool. In some cases the tool sweep volume may be modeled using primitives. The finished part is then:

$$C = C_1 - \sum_{i=1}^{n} V_i$$

where n is the number of tool motions. Since this model only uses a difference operator, the model can also be called a Destructive (or Deforming) Solid Geometry (DSG) model [Arbab 1982]. It is a subset of the more general constructive solid geometry. When a part is designed using DSG model, there is an analogy between the modeling sequence and the machining sequence.

2.3 Boundary Models

A boundary model (B-Rep) represents an object by its bounding faces (Fig. 2.18). A vector (or orderly list of edges) on each face points to the inside of the solid; therefore, the inside and outside of the object can be told. The topology is preserved in a tree. There is no operator in the tree, and the nodes of a tree are faces, edges, and vertices. The terminals of the tree are geometrical entities and normally vertices. Since on a face there may be more than one loop of edges, it is common to add another entity, loop, in the B-Rep. For example, Fig. 2.19 shows a face with three loops, an outer loop of straight edges and two inner loops of a circle. A loop is an orderly list of edges (Fig. 2.20). The direction of the inside of the object may be defined by the order of the edges, i.e., use right hand rule; thumb points to the material side of the face.

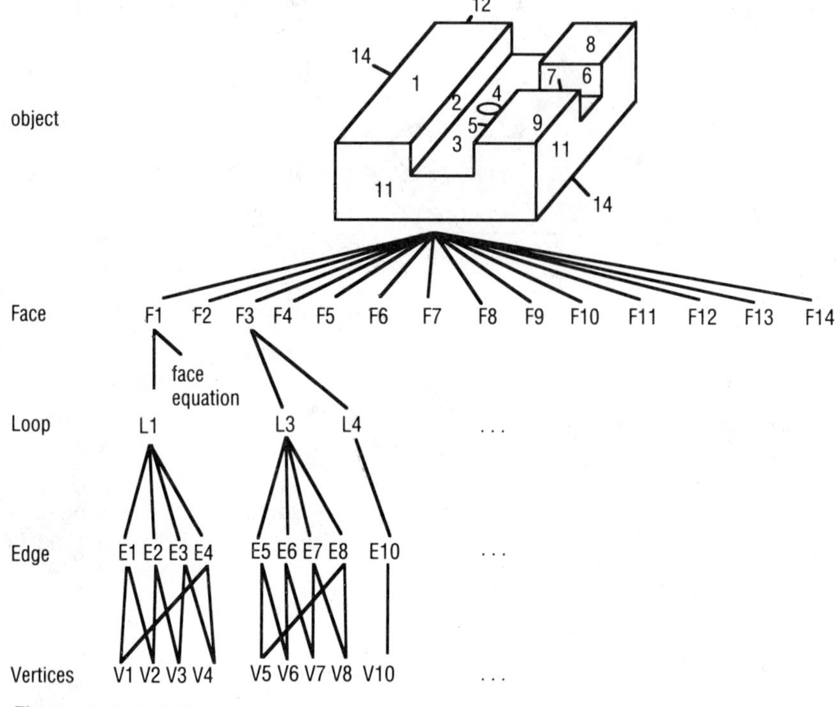

Figure 2.18 A B-Rep

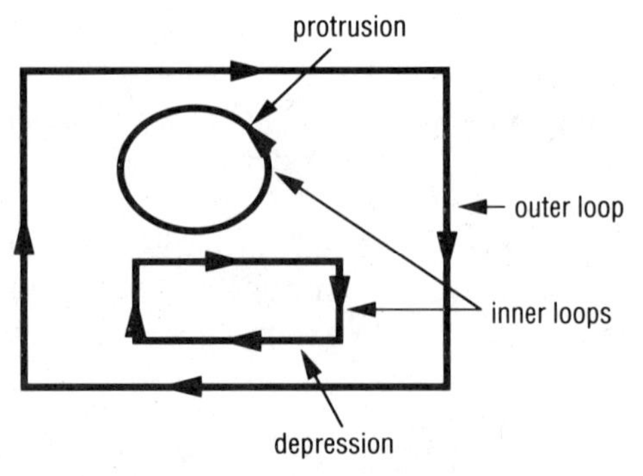

Figure 2.19 Inner and outer loops

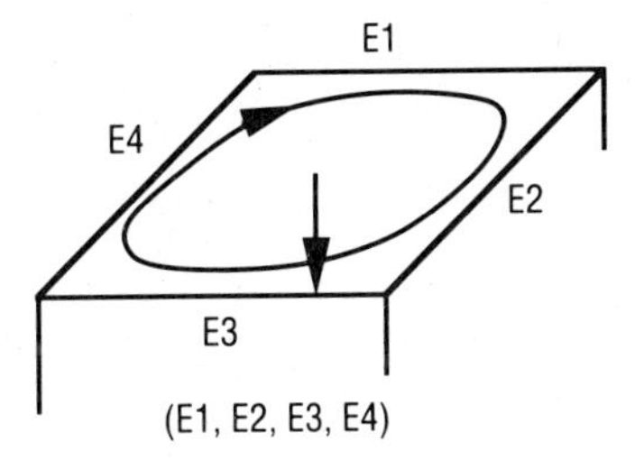

Figure 2.20 Face loop

Since a B-Rep represents faces, edges, and vertices explicitly, it is much easier to display a model. To display the wireframe of an object, one can simply traverse the edges and plot each one of them. The existence of faces enables the graphics system to do hidden surface removal; thus, a better picture of the model can be displayed. For NC application, the most important information is face and edge definition. One can again use this data for interactive NC part programming.

However, B-Rep is not without its shortcomings. A B-Rep model usually is much larger than the corresponding CSG model. It requires a large amount of memory to store. The model is difficult to build directly from faces, edges, and vertices. The model is also not unique (Fig. 2.21). All four models in Fig. 2.21 represent the same object. A user can legally build any of the four models (there are actually an infinite number of models) for the same object. However, the same non-uniqueness problem exists in the CSG model.

The validity of a B-Rep model can be checked using Euler's formula. In actual implementation, care is taken by the modeler to ensure that the solid generated preserves Euler's equation. Euler's

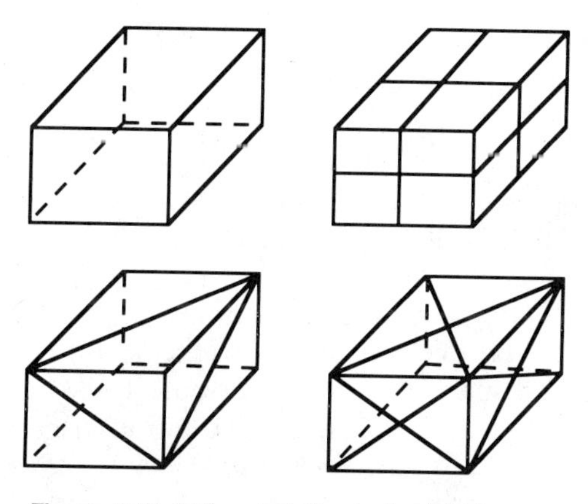

Figure 2.21 Different B-Reps of a block

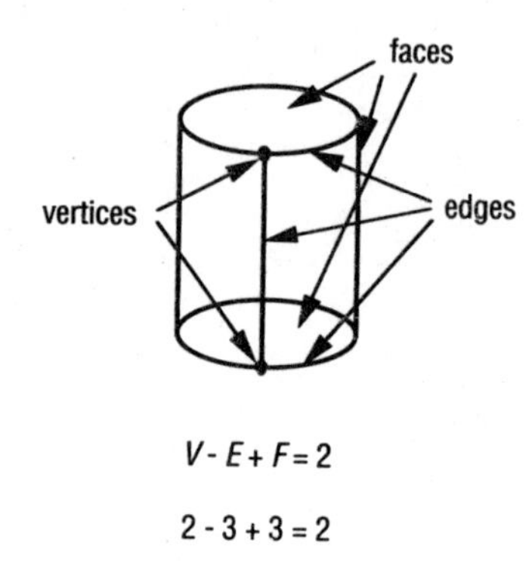

$$V - E + F = 2$$

$$2 - 3 + 3 = 2$$

Figure 2.22 A cylinder and validity check using Euler's formula

formula states the following: For any topologically valid solid the number of vertices plus the number of faces is equal to the number of edges plus two.

$$V - E + F = 2$$

Where V is the number of vertices, E is the number of edges, and F is the number of faces. The formal proof is not given here. The reader may try to apply this formula to the objects shown in Fig. 2.21. Figure 2.22 shows a cylinder and how to apply the formula to it. It is obvious that this formula does not apply to the object in Fig. 2.18. To make it true, the following conditions must also be true:

- No hole is present.
- The solid is connected.
- Each edge is adjacent to exactly two faces and terminated by two vertices.
- Each vertex is the intersection of at least three faces.

When there are holes in an object, the formula is modified:

$$V - E + F - H + 2P = 2$$

where H is the number of inner loops, and P is the number of passages (through holes). For example, for the object in Fig. 2.18, $V = 26$, $E = 38$, $F = 14$, $H = 2$, $P = 1$, and $S = 1$. Therefore, $26 - 38 + 14 - 2 + 2 = 2$, the object is topologically valid. A set of Euler's operator may be defined to construct the B-Rep of an object. These operators must maintain Euler's formula to be true and must be as simple as possible. For example, adding an edge and a vertex will add one to V and add one to E, the effect is cancelled in the formula. There is no guarantee that the intermediate object is a solid. However, the completed model should agree with the formula.

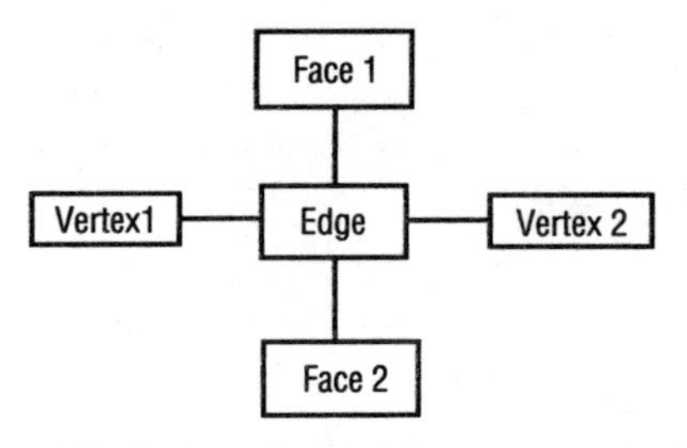

Figure 2.23 A winged edge data structure for B-Rep

As mentioned earlier, usually a B-Rep model is constructed using a boolean operator front end. The model is then evaluated into a B-Rep model. For example, the system ROMULUS (by Shape Data of UK, a subsidiary of Evens and Sutherland, now it is called Para Solid and marketed by McDonnell Douglas) is a B-Rep-based system. A designer would use boolean operators and solid primitives to build a model. The model is then evaluated into a B-Rep model. In ROMULUS, all subsequent graphics operations or engineering computations use the B-Rep. When this is the method, a B-Rep is constructed, regardless of the sequence a model is constructed, and the final B-Rep should be the same. Uniqueness may not be a problem. The non-unique CSG models can actually be evaluated into an unique B-Rep. This may be a good reason for using B-Rep in feature recognition.

In actual implementation, a B-Rep model data structure is of the network type. A commonly used data structure is called a winged edge data structure (Fig. 2.23) [Eastman 1979]. A topologically valid B-rep always has each edge adjacent to exactly two faces and terminated by two vertices. A winged edge data structure has an edge as the center, it links to exactly two vertices and two faces. In this data structure the edge is the center of the data structure. Since each edge is adjacent to exactly two vertices (in circle, the two vertices happens to be identical; this can be taken as a special case) and two faces, this data structure forms a basic element in a B-Rep tree. The linkages between each data structure elements are pointers. Given any edge, it is simple to find the adjacent faces and vertices.

2.4 Feature-based Design

The last but not the least important design representation scheme is feature-based design. Feature-based design means a design with features. Unfortunately, at the moment, the term *feature* does not have a definition which is agreed upon by everyone. It is definitely application specific. Webster's New Collegiate Dictionary defines feature as: "1. the structure, form, or appearance esp. of a person, 2. the makeup or appearance of the face or its parts, 3. a prominent part or characteristic." Used here, feature refers to "a subset of geometry on an engineering part which has a special design or manufacturing characteristic." In this book, what we are interested in is the form (geometry) feature and not the functional feature.

Feature-based design is used for the following reasons [Rossignac 1988]:

- capture designer's intension
- simplify the specification of the part's geometry
- provide a concise description of the part's characteristics

In a feature-based design system, a designer uses some features to define an engineering object. In this chapter, we will limit the engineering object to engineering parts. A feature has its specific geometry and must be associated with some feature attributes. The attributes can be dimensions, dimensional tolerance, manufacturing notes, etc. Depending on the application, different information may be included. In any case, a feature is a geometrically independent entity. It contains some meanings useful to the application it is designed for. It is also the building block for an engineering part.

Features can be classified by a few different ways. Based on the geometry, one may classify features into the following:

- face features—features defined by two or three dimensional faces.
- volumetric features—features defined by three dimensional, enclosed volumes.

Based on the applications:

- design features (DF)—features meaningful to design.
- manufacturing features (MF)—features meaningful to manufacturing.

For 2D or surface modeling CAD system, face features may be used. Some typical features can be hole pattern, gear, fillet, hexagon, etc. Volumetric features are best used in conjunction with a CSG modeler. The features are primitives which carry special meanings. Set operators can be used to construct a solid model using features and regular solid primitives. Some typical volumetric features are hole, boss, simple slot, T slot, V slot, pocket, keyway, groove, slab, notch, profile, cutout, etc.

Design features and manufacturing features can both be either face features or volumetric features. Design features are features which are meaningful to the design. Some design features are hole, chamfer, groove, countersink, counterbore, taper, screw thread, etc. A list of form features can be found in Allen [1980]. Manufacturing features are those features which are meaningful to the manufacturing. Some manufacturing features are hole, groove, countersink, counterbore, pocket, hole tip, chamfer, fillet, etc. Actually design features and manufacturing features do overlap ($DF \cap MF \neq \emptyset$). From the list above, one can find that many features are identical. However, some are different, and some of them even use the same name but carry different meaning. For example, hole tip is a manufacturing feature but not a design feature. The designer can care less about the tip; it is just something extra to a hole. Since drilling always creates a hole tip, a manufacturing engineer will have a hard time if no tip is specified.

Some of the advantages of using a feature-based design system for engineering design has been mentioned. Many design features and manufacturing features are similar. Sometimes there is a one-to-one correspondence of the two. A part designed using features may be planned automatically.

Feature-based design is not without some deficiencies. Some of them are discussed below. When used in a real world design environment, too many features need to be considered. A part cannot be designed with feature primitives only; it needs intermediate surfaces to connect feature primitives. It is difficult to mix them in a modeling system. Design features and manufacturing features are not always the same. Process planning and NC generation is not always straightforward. It is easy to use for simple parts and extremely difficult to use for complex parts (same as that in a CSG modeler).

To develop a feature-based design system for process planning, the following factors must be considered:

- system should allow user to design new features
- features should contain a method which defines the characteristics of the feature
- in order to use it for process planning, further geometric reasoning must be done

A feature based modeler for design and process planning is presented in Chapter 6. In such a system, the process planning system requires an analysis of the geometry of the finished feature model. Geometric reasoning is an important part of the integrated system.

2.5 Conclusions

In this chapter several part-modeling methods are discussed. The emphasis is on the methods which have been used or can potentially be used in an automated process planning system. From the discussion, we may conclude that the traditional design representation methods are appropriate for certain types of design work. We will not see them go away for a long time. However, to use them directly in an automated process planning system is not possible. What has the most potential are representations which are complete and unambiguous, such as solid models and feature-based models. However, the latter are not without their own difficulties. They are limited in the domain of geometry they can model. When used to model complex parts, the modeling task can be more difficult than that of a more traditional approach. Currently, solid modelers also lack the tolerancing capability. Although a lot of effort is being spent on improving these weaknesses, the efficient and effective method has yet to be developed.

Since a process planning system is driven by the input data, it is extremely important to select a representation which is most suitable to the application. Depending on the complexity of the part to be planned, the degree of automation sought, and the nature of the design process, a representation can be selected. If the current design practice cannot be changed, then, either a manual or an automated interface should be developed to prepare the data for the process planning system. Otherwise, a new design system which takes into consideration the need of manufacturing planning should be constructed. Since process planning is feature-based, it is most appropriate to use a feature-based model as the internal data representation. Again, the feature may be two dimensional or three dimensional, depending on the application domain. In any case, the features are manufacturing features which may not be the same as the design features. How to integrate the design with the rest of the process planning functions is the most important consideration.

In the next chapter, automated CAD interface will be discussed, as will several approaches which extract manufacturing features from a solid model. Chapter 6 presents a system, QTC, which integrates a feature-based design system with an expert planning system. It offers a simple solution in a limited domain for a complex problem.

References

Allen, D., "A glossary of form features," Report, Computer-Aided Manufacturing Laboratory, Brigham Young University, Provo, Utah, 1980.

ANSI Standard Y14.5-1973, Dimensioning and Tolerancing, United States of America Standards Institute, N.Y., 1973.

Arbab, F., "Requirements and architecture of a CAM oriented CAD system for design and manufacture of mechanical parts," Ph.D. Thesis, University of California, Los Angeles, 1982.

Baer, A., Eastman, C., and Henrion, M., "Geometric modelling: a survey," *Computer-Aided Design,* vol II, no. 5, September 1979.

Bezier, P., *Numerical Control: Mathematics and Applications,* Wiley, 1972.

Braid, I.C., "Designing with Volumes," CAD Group, Cambridge University, Cambridge, England, 1973.

Chang, T.C., and Wysk, R.A., *An Introduction to Automated Process Planning Systems,* Prentice-Hall, Englewood Cliffs, N.J., 1985.

Chang, T.C., Anderson, D.C., and Mitchell, O.R., "QTC—An integrated design/manufacturing/vision inspection system for prismatic part," Proceedings of the ASME 1988 Computers in Engineering Conference, Volume 1, San Fransisco, Calif., July 31–August 3, 1988, pp 417–426.

Coons, S.A., "Surfaces for computer aided design of space forms," Report number MAC-TR-41, Project MAC, M.I.T., 1967.

Ding, Q. and Davies, B.J., *Surface Engineering Gemometry for Computer-Aided Design and Manufacture,* Ellis Horwood, 1987.

Dixon, J.R., Cunningham, J.J., and Simmons, M.K., "Research in Design with Features," IFIP WG 5.2 First International Workshop on Intelligent CAD, Cambridge, Mass., October 6–8, 1987.

Eastman, C., and Weiler, K., "Geometric modeling using euler operators," Proceedings of First Annual Conference on Computer Graphics in CAD/CAM Systems, MIT, April 1979, pp. 248–254.

Faux, I.D., and Pratt, M.J., *Computational Geometry for Design and Manufacture,* Ellis Horwood, 1979.

Ferguson, J.C., "Multivariable curve interpolation," *Journal ACM,* vol 11, no. 2, 1964, pp. 221–228.

Henderson, M., "Automated group technology part coding from a three-dimensional CAD database," Bound Volume, Symposium on Knowledge Based Expert Systems for Manufacturing, ASME Winter Annual Meeting, Anaheim, Calif., December 7–12, 1986, pp. 195–204.

Johnson, R.H., "Dimensioning and tolerancing final report," Rep R-84-GM-02.2, Computer Aided Manufacturing-International, Inc., Arlington, Tex., May 1985.

Joshi, S., Vissa, N.N., and Chang, T.C., "Expert process planning system with solid model interface," *International Journal of Production Research,* vol. 26, no. 5, 1988, pp. 863–885.

Kung, H., An investigation into development of process plans from solid geometric modeling representation, Ph.D. Thesis, Oklahoma State University, 1984.

Kyprianou, L.K., "Shape classification in computer aided design," Ph.D. Thesis, University of Cambridge, Cambridge, U.K., 1980.

Mortensen, M.E., *Geometric Modeling*, John Wiley, 1985.

Okino, N., and Kubo, H., "Technical information system for computer-aided design, drawing, and manufacturing," Proceedings of the Second PROLAMAT 73, 1973.

Requicha, A.A.G., and Chen, S.C., "Representation of geometric features, tolerances and attributes in solid modellers based on constructive geometry," *IEEE Journal of Robotics and Automation*, vol RA-2, no. 3, September 1986, pp. 156–166.

Requicha, A.A.G., and Voelcker, H.B., "Solid modelling: a historical summary and contemporary assessment," *IEEE Computer Graphics and Applications*, vol 2, no. 2, March 1982, pp. 9–24.

Requicha, A.A.G., "Mathematical models of rigid solid objects," Technical Memo No. 28, Production Automation Project, University of Rochester, Rochester, N.Y., 1977.

Requicha, A.A.G., "Representations of Rigid Solid Objects," Computer Aided Design Modeling, Systems Engineering, CAD Systems, (ed. Encarnacao, J.) CREST Advanced Course, Darmstadt, September 1980, Springer-Verlag, Berlin.

Requicha, A.A.G., "Solid modeling: current status and research directions," *IEEE Computer Graphics and Applications*, vol 3, no. 7, October 1983, pp. 25–37.

Rossignac, J., "Features in geometric design," UCLA/NSF Workshop on Features in Design and Manufacturing, February 26–28, 1988.

Turner, G.P., and Anderson, D.C., "An object oriented approach to interactive feature-based design for quick turnaround manufacturing," Proceedings, ASME Computers in Engineering Conference, Vol 1, San Fransisco, Calif., July 31–August 3, 1988.

Voelcker, H.B., and Requicha, A.A.G., "Geometric modeling of mechanical parts and processes," *Computer*, vol 10, no. 48, 1977.

Wang, H.P., and Wysk, R.A., "Micro-GEPPS—a microcomputer based process planning system," Bound Volume, Symposium on Computer-Aided/Intelligent Process Planning, ASME Winter Annual Meeting, Miami Beach, Fla., November 17–22, 1985, pp. 139–150.

3

Design Interface for Process Planning Input

Introduction

The first step in process planning is to understand the engineering design. As discussed in the previous chapter, an engineering design can be represented in several different ways. For a human being the easiest representation to understand is a drawing, as they say: "A picture worth a thousand words." For an untrained person, a three dimensional drawing provides the most direct representation of an object. From this perspective, the overall geometry of an object can be easily identified. However, the information obtained from such a representation is not precise enough. Detailed geometric information and dimensions cannot be obtained. For manufacturing purposes, a detailed and precise model of the object is necessary. Without it one won't be able to duplicate the designer's intention on a piece of raw material. For this reason, engineers prefer to use three-view (three dimensional) engineering drawings. In such a drawing, details of an object are not only represented by geometries, but they are also supplemented with drafting symbols and texts. The design details can thus be better interpreted.

When performing discrete part manufacturing process planning, a process planner usually first identifies the overall shape of the part. Such information is crucial for deciding the way a part is to be processed. The overall shape can be recognized from reconstructing the contours of the three projected views. It is difficult for most of us to explain why we view a drawing as a turned part or prismatic part. With only a few drafting rules in mind, most engineers based on their experience are able to reason out the shape of the part in a drawing. The spatial reasoning capability of a human being is still a mystery to us. We seem to be able to pick from among a large number of data, the necessary ones to deduce a conclusion. After the overall shape of a part is found, manufacturing features, datum surfaces, and additional manufacturing information are recognized and extracted from the drawing. The task of transforming a design drawing into manufacturing information is called design interface.

This is probably the most important and most difficult interface task in process planning. In the early days of automated process planning no one attempted to complicate the already difficult task of process selection by considering automated design interface. There are several ways to get around this problem. Most of them involve the use of humans to serve as the interface. Normally a drawing is translated into the process planning specific data (such as features) by a human user. One of the prototype systems [Chang 1982] used interactive interface information required by a process planning system and is interactively identifiable by a human user. These approaches are sometimes very tedious. The final process plan is greatly influenced by the decisions of the human user. Therefore, an automated interface seems to be desirable.

There are two major tasks involved in the design/process planning interface: model decomposition and feature recognition. The decomposition task separates the features from a part model, while the recognition task classifies and identifies the semantics of the feature. An experienced human being usually performs these two tasks intermittently. A potential feature is spot-based on some visual clue, such as a circle on a face, a step curve, etc. A guess is made, then more geometrical entities (in a drawing, lines and curves from different views) are collected into the feature. If this approach fails, one tries other elements or makes another guess. Finally, when a feature is clearly identified, the geometrical entities included in this feature are removed from the picture. The removal of geometrical entities is a decomposition operation. This process is repeated for another feature, until all features are removed. Of course, most of the time the geometrical entities are not removed physically from the drawing or picture. They are collected, however, into different feature sets stored in our mind.

When looking closely enough, one may find that the automated interface problem is a problem of recognizing manufacturing features from a design. As difficult as it seems to be, automated picture understanding is by no means new. One of the major fields of artificial intelligence, machine vision, studies this subject: "Vision is the information-processing task of understanding a scene from its projected images," [Chapter XIII, Cohen and Feigenbaum 1982]. There are three major fields in machine vision: signal processing, classification, and understanding. The classification and the understanding of pictures (design models) are related to the design interface. The general method used in classification is called pattern recognition. In vision, understanding a scene usually implies the interpretation of a two dimensional picture into three dimensional meaning. Pattern recognition focuses mainly on the understanding of a two dimensional pictures. In the design interface problem, although less desirable, a "picture" can be a two dimensional drawing, or a three dimensional geometric model. The information needed for process planning is feature information. The problem is to recognize features from a design representation.

Since the purpose of this chapter is to discuss the various ways of recognizing features from a design representation, we will base our discussion on the historical development of different approaches. As mentioned before, the feature recognition problem can trace its roots to machine vision in Artificial Intelligence. In machine vision studies, the input is an image of a picture. This picture is first processed into machine-understandable digital format, a series of ones and zeros. Patterns (lines, curves, regions, etc.) are then recognized from the digital data (image). The semantic information of the patterns is then extracted. The pioneer work can be traced to Roberts [1965]. In the study he dealt with images consisting of blocks. Engineering design models are very well defined; therefore, the first step—image processing—is unnecessary for design interface. Even the second step, pattern recognition can be eliminated since the features in the design are defined slightly different from the ones defined in machine vision. The major influence of vision to design interface lies on the syntactic pattern recognition methods.

The syntactic pattern recognition [Fu 1974, Rosenfeld 1979] method uses language theory for scene analysis. Patterns are defined formally with pattern grammars. A picture is then represented by a concatenation of pattern primitives. Patterns are also represented by such primitives. A parser is used to identify whether a given piece of picture satisfies a certain grammar, thus having a certain pattern. This approach is also called automata. It also has a root in formal language theory. Several works in design interface using syntactic pattern recognition methods can be found [Kyprianou 1980, Staley, Henderson, and Anderson 1983, Liu and Srinivasan 1984, Choi 1983]. Kyprianou used the boundary representation of a part to recognize the depressions (such as holes) and protrusions (such as boss). A rule-based system uses the result to produce a Group Technology classification code. Staley, et al. developed pattern grammars for different holes (represented by cross sections). Given a string representing a hole using picture primitives, a parser can determine which hole class the hole belongs to. Srinivasan and Liu proposed to represent the geometrical capability of a machining center using pattern grammar. Their pattern grammar also uses the cross section and assumes the symmetry of the pattern. Choi represents the two dimensional projection of a hole using picture primitives. Cutting tools are also represented in the same way. A Pascal program is written to match the cutting tools with the holes since the syntactic pattern recognition approach can only recognize a feature when a unknown one is given. Most of the existing works based on syntactic pattern recognition approach can only recognize a feature when an unknown one is given. Most of the existing works based on syntactic

There are also some other approaches used in design interface. Some of these approaches recognize feature elements from a given solid model of a part. Others may decompose a solid model into unknown primitives. Three approaches can be identified: geometry decomposition, logic, and graphic theory. The geometry decomposition approach [Woo 1982, Grayer 1977] decomposes the original solid model into smaller solid objects. In Woo [1982] an alternative series decomposition is used. The results of the decomposition can either be a series of additive (assembly) or subtractive (machining) solids. There is no way of controlling the decomposition to produce meaningful solid objects. Grayer's work emphasizes the generation of NC tapes. The difference between a raw material and the final part is decomposed into a set of disjointed laminae. A cutter path is independently generated for each laminae. The above two works do not, however, provide semantic information of the features.

The logic approach is associated with an expert system. The topological structure of a feature is written into logic rules. The input to the system is usually a boundary model. For example, a round blind hole can be described as an entity with a inner circular loop connected to a cylindrical body and a conical bottom. Henderson [1984] developed feature rules for holes, pockets, and slots. Implemented in a logic programming language—Prolog—the system can search through the data structure of a boundary model to extract semantic knowledge. Both the decomposition and recognition operations are accomplished. Kung [1984] developed an expert system for extracting disjointed features.

Most of the above methods are computational expansive and limited to $2^1/_2$ D objects. Joshi [1987] used a heuristic approach to simplify the complexity of the problem. First the boundary representation of a part is represented using an attributed adjacency graph. Heuristics are used to remove most of the unimportant nodes and links from the graph. The remaining subgraphs are used for feature recognition. Features are also classified in a hierarchical form. The recognition is done through several stages, the global classification and then the fine classification. Features are recognized then extracted from the solid model data structure.

To date, none of the methods developed is perfect and all are far from ready to be used in commercial systems. A great deal of research work is still needed in order to complete the work. In the following sections, several of the existing approaches will be discussed in detail.

3.1 Syntactic Pattern Recognition Approach

3.1.1 Introduction

In syntactic pattern recognition, a picture is represented by some semantic primitives, which are written in a picture language. A set of grammars consisting of some re-write rules define a particular pattern. A parser is then used to apply the grammar to the picture. If the syntax of the picture language agrees with the grammar, then the picture can be classified as belonging to the particular pattern class. This is very similar to nature and formal language processing in which a sentence can be analyzed to see whether it is grammatically correct. Similarly, a statement written in a computer language is parsed to see whether it posses correct syntax. In any of the compilers, a parser is always used first to check the syntax of the user written program. The same principle can be applied to design interface. Let us use a simple design as an example. Say, in a two dimensional drawing, a hole is represented by its projection (Fig. 3.1).

There are four pattern primitives defined in Fig. 3.2, they are A, B, C, and D. Each one of them has a unit length. We can substitute a line drawing with the pattern primitives. The drawing in Fig. 3.1 can be represented by a string "BBCCCBBAAABB." Notice that we arbitrarily selected to start the drawing from the left. A pattern

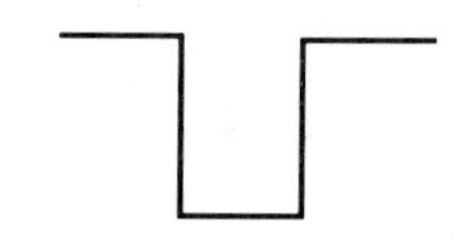

Figure 3.1 A simple hole drawing

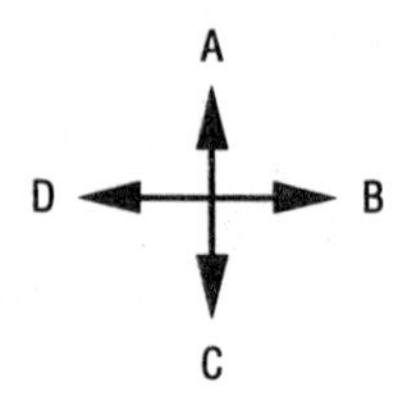

Figure 3.2 Pattern primitives (not on scale)

grammar can then be written for the simple holes. The simple hole class is a language in the form of "$C^nB^mA^n$" where C^n denotes n segments of primitive Cs. The grammar can be formally defined by a four tuple:

$$G = (V_n, V_t, P, s).$$

where

$V_t = \{A, B, C\}$ /* is a set of terminal symbols
/* (pattern primitives).
$V_n = \{s, a, b\}$ /* V_n is a set of non-terminal
/* symbols: starting symbol and variables.
/* s is a starting node, and a is a variable.
$P = \{s \Rightarrow CaA$ /* P is a set of production rules.
$a \Rightarrow b|\ CaA$ /* a vertical bar denotes OR
$b \Rightarrow B|\ Bb\ \}$

The language $L(G) = \{ \quad C^nB^mA^n \mid m,n = 1,2,\ldots\}$

A parser is needed to parse the string "BBCCCBBAAABB" using the given grammar. We will discuss the parser later; first let us see how the production works. Beginning with the starting symbol s.

$s \Rightarrow CaA$ /* apply the first rule
$\Rightarrow CCaAA$ /* apply the second part of the second rule
$\Rightarrow CCbAA$ /* apply the first part of the second rule
$\Rightarrow CCBAA$ /* apply the third rule

Since the string consists only of the terminal rules, no further production can be made. We can conclude, then, that string "CCBAA" represents a simple hole. In this example, we showed the pattern generation process. We can apply this grammar to generate an infinite number of simple holes. All these simple holes will be similarly defined except that they will be of different sizes. The pattern classification process is the reverse of the hole generation process. Before we comment more about the use of the syntactic pattern recognition method for feature recognition, let us have a quick overview of the general method.

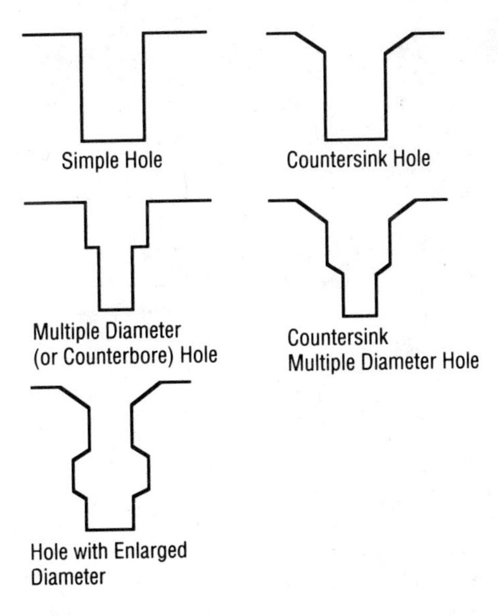

Figure 3.4 Hole classes can be defined

A three dimensional part model can be projected onto a plane. The scope of features recognizable are linear and rotational sweep features (holes, and 2½ D features). The projection is the generating curve and the projection direction is the sweeping direction. Figure 3.6 shows several sweeping features and their pattern descriptions.

Feature grammars are prepared for each feature. For example, the hole with an enlarged diameter in Fig. 3.4 can be written in a language as follows:

$$L(G) = \{ D^{n1} E^{n2} F^{n3} E^{n4} D^{n5} E^{n6} C^{n7} A^{n6} B^{n5} A^{n4} H^{n3} A^{n2} B^{n1} \mid {}_{n1,n2,n3,n4,n5,n6} = 1 \text{ to } n\}$$

Since the production rules for the grammar are lengthy, they will not be presented here. Similarly, the dovetail slot can be written as:

$$L(G) - \{F^{n} C^{m} H^{n} \mid {}_{n,\ m,} = 1 \text{ to } N\}$$

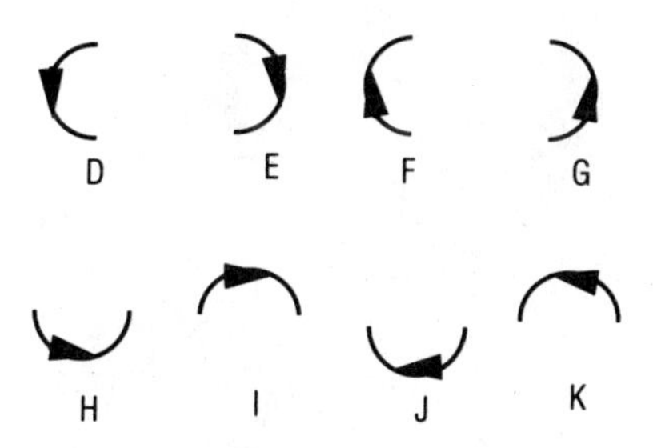

Figure 3.5 Additional pattern primitives

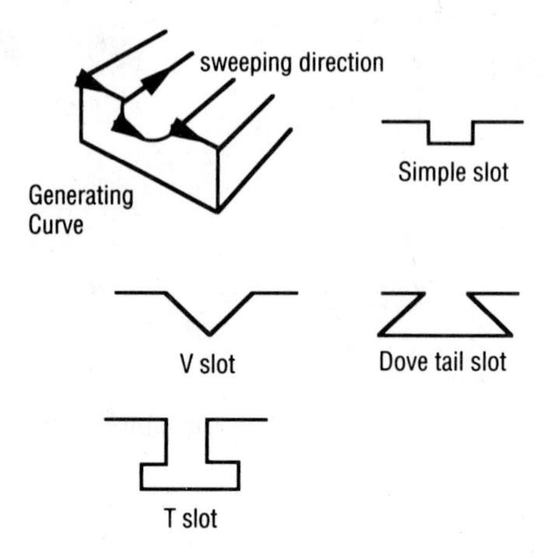

Figure 3.6 Some sweeping features

It can be seen that when this method is taken, it is difficult to distinguish between a simple hole and a simple slot. Additional information is needed for further classification. The structure of the system for feature recognition is shown in Fig. 3.7. An unknown feature has to be parsed using each grammar. Whichever grammar it can pass through, the feature is thus classified. When there are many feature classes to be classified, the process can become very lengthy.

Liu and Srinivasan [1984] proposed to define the shape producing capabilities of a machining center using pattern grammars. For example, all the features producible by a machining center can be modelled using a pattern grammar. Certain production rules are used to model the shape capability of a machining process. In a V slot case two production rules can be written: $\{S \Rightarrow DaB\}$ and $\{a \Rightarrow a|DB\}$. If these two rules are used then a dovetail tool is needed. In the system Liu and Srinivasan proposed, rules are marked as certain process generating rules. They proposed that not only can machining centers be selected, but the process plans can also be generated. In a

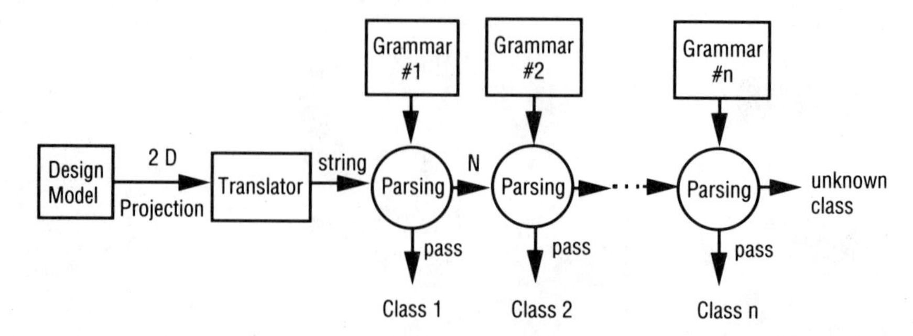

Figure 3.7 An architecture for a syntactic pattern recognition based feature recognizer

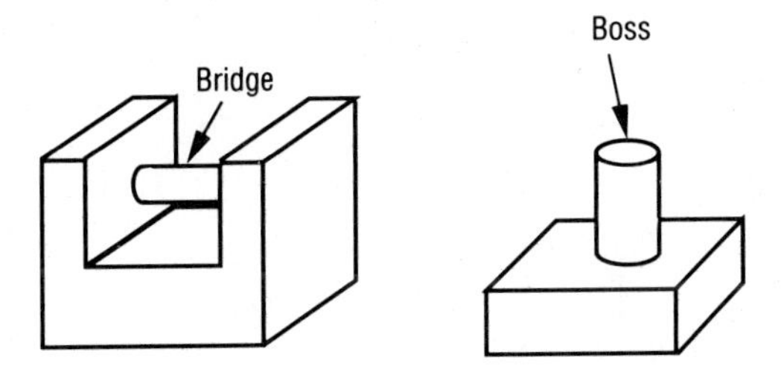

Figure 3.8 Protrusions: bridge and boss

later work Srinivasan and Liu [1987] discussed using tree grammar to represent machine capabilities.

Kyprianou [1980, see also Jared 1984] used the syntactic pattern recognition method for automated group technology coding and classification. In his work, he first recognized the three dimensional features from the solid model of a part. The features recognizable are depressions, such as slots, holes, and pockets; and protrusions, such as bosses and bridges (Fig. 3.8). In order to recognize three dimensional features, he used structural primitives instead of simple geometrical features. The input to his system is a boundary representation generated by the Build solid modeler. In order to make the recognition efficient, a boundary model is first decomposed into smaller objects (facesets). The relationships between facesets are identified and saved for later use. Each faceset is then recognized using the pattern recognition method. The recognized features with the relations are used in a rule-based system for GT code generation.

One natural way to decompose a boundary model is to cut the inner loops from a face. An inner loop in a face is created by either a depression or a protrusion (Fig. 3.9). In the figure the two loops, loop1 and loop2, are both associated with face F1. The solid model data structure is designed in such a way that inner loops are identified. F1 is called the primary face. One can also find the adjacency relations between F1 and the faces containing the edges in the inner loop. If they are convex, then the relation between the root faceset (include F1) and the leaf faceset (the one expanded from the inner loop) is convex. The faces linked with the primary face are collected into one faceset, and the faces linked to the inner loop are collected into another faceset. When a face linked with the inner loop is not found then two other rules can be used to find the primary face, (1) faces which contain at least one concave edge, or (2) a curved concave face. The relations between two facesets can be convex, concave, and hybrid (partially convex and partially concave).

Only after facesets are decomposed, can feature recognition start. In the recognition stage, the structural primitives used are defined by

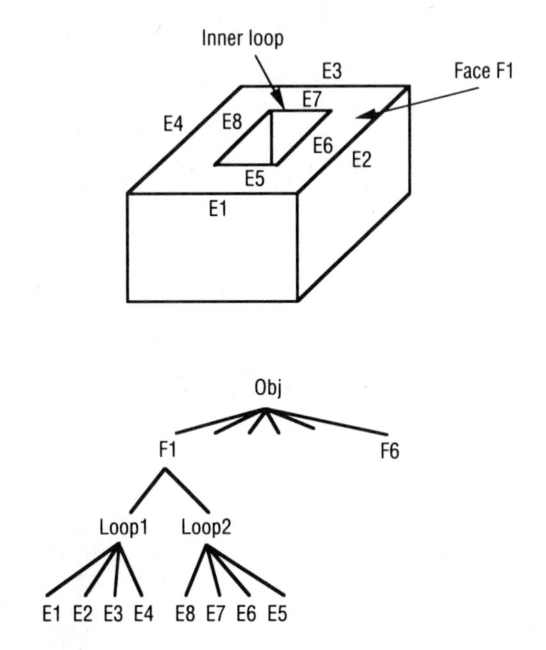

Figure 3.9 Boundary representation of a hole in an object

face adjacency relations, they are concave adjacency, convex adjacency, and smooth adjacency (Fig. 3.10). The structural information can be derived from the faceset directly.

The grammars used for recognition is similar to that discussed before. For example, a grammar for a noncylindrical hole can be written pictorially as it is in Fig. 3.11. The symbol with dashed line implies a third loop and is in between the current loop and the selected adjacent loop. There is no edge at the dashed line. This grammar can produce noncylindrical holes with a smooth surface or an intersecting geometry. The parsing of a hole is shown in Fig. 3.12.

With the above semantics of individual features and their relations known, the next step is to generate a GT classification code. In a GT code digit, each value has its special meaning. Not all special mean-

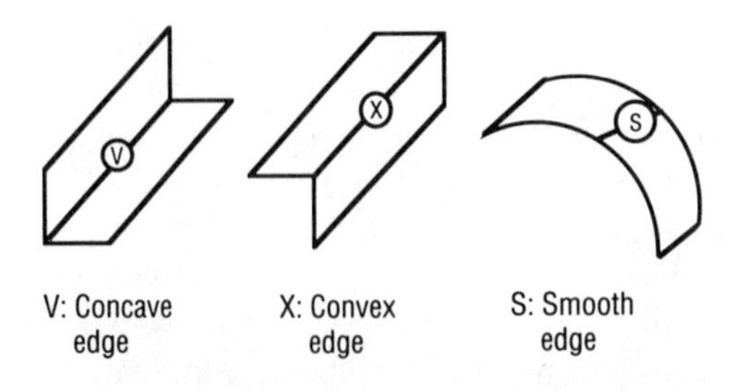

Figure 3.10 Structural primitives

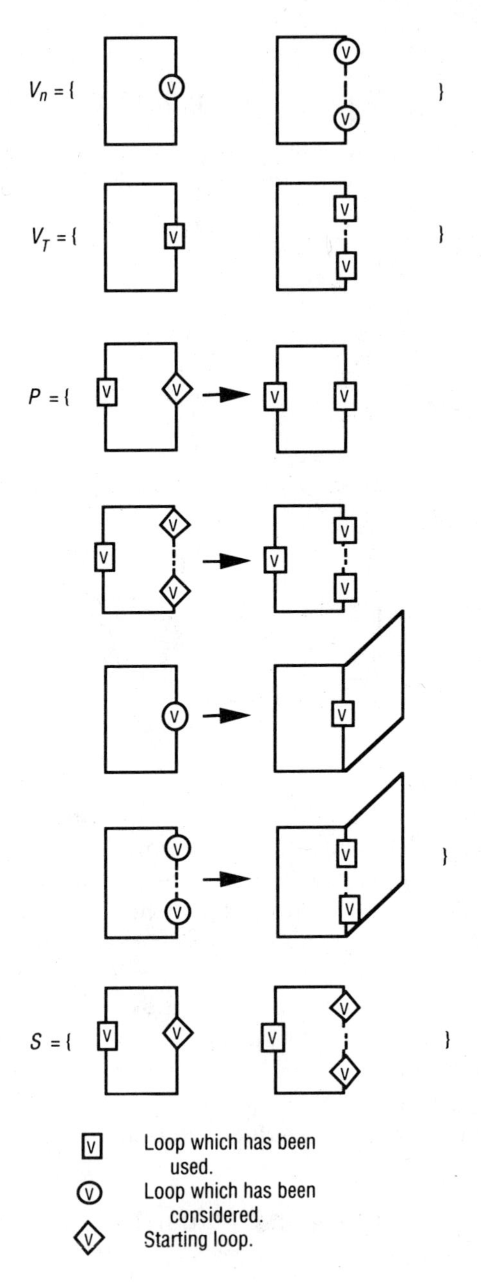

Figure 3.11 A grammar for noncylindrical holes

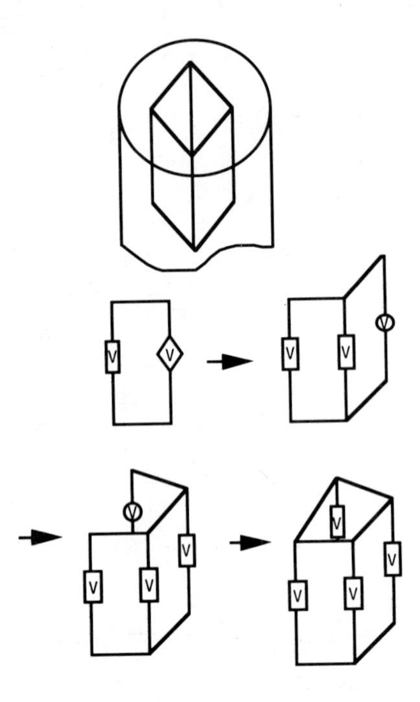

Figure 3.12 Parsing a noncylindrical hole

ings can be determined through syntactic structure, however. Therefore, syntactic pattern recognition is not appropriate for this application. This application can be done, however, using a rule base approach. A cylindrical component can be identified by a part with no non-cylindrical external feature. This can be written in a simple rule. The rule will use a procedure which checks the feature types. A step leading to the shape of both ends can be defined as a cylinder connected to cylindrical bosses at both ends. Since only simple features (depressions and protrusions) can be recognized, the code digit on gear teeth, planner surface machining, etc., cannot be identified. The code digit is related to dimension, tolerance, and manufacturing information, which is needed for additional processing.

3.2 State Transition Diagram and Automata

In a system called CIMS/PRO [Iwata et al. 1980], a similar idea was presented where a part is described using a system called CIMS/DEC [Kakino et al. 1979]. In this system, the part geometry is described by sweeping operations, and/or the union of the sweeping volumes. The generating surface is described by ordered pattern primitives together with technological information and approach directions. The machining feature on the generating surface is recognized using a

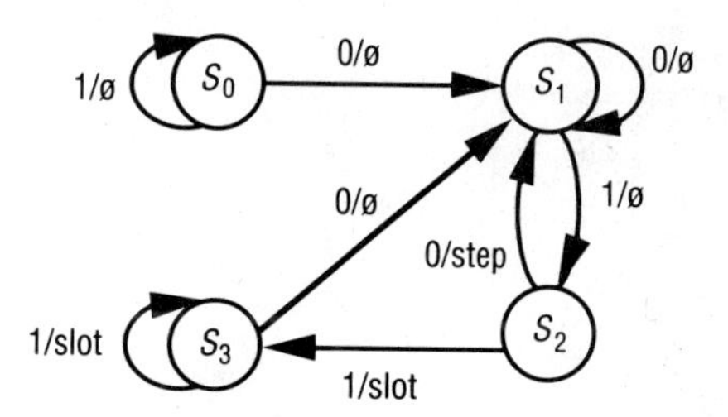

Figure 3.13 State transition diagram for step and slot

state transition diagram. Instead of using grammars and primitives, the relationships of adjacent pattern primitives are used. Convex adjacency is assigned a value of "0" and concave adjacency is assigned a value of "1." A state transition diagram for step and slot is shown in Fig. 3.13. In the state transition diagram, S_i, i = 0 to 3, are states. The term a/b means input/output. For example, 1/slot, means if the input is 1, output the feature class as slot and go to the next state. The symbol ø means no output.

The simple slot in Fig. 3.6 can be represented by a string "0110." Using the state transition diagram, the following operation is taken:

Input	Output	State
		S_0
0	ø	S_1
1	ø	S_2
1	Slot	S_3
0	ø	S_1

Within the class, classification is then done by using the feasible approach directions. For example, a slot with three feasible approach directions is an open slot. The one with two feasible approach directions is a closed slot. In a degenerated case, when only one feasible approach direction is available, the slot is called a pocket. Since only the adjacency information (0,1) and feasible approach directions are used for recognition, the system does not further recognize whether a slot is a simple slot or a dovetail slot. One can easily extend the work by checking the angles between two adjacent primitives. Further discussion can be found in the section on the graph-based approach on feature recognition.

The state transition diagram is similar to the context-free grammar we discussed. If we substitute the states with non-terminal symbols, and input by the terminal symbols (0 and 1), a grammar can be written. In Milacic's work discussed below, this method is also used.

Milacic [1985] discussed the use of automata for part family classification. Grammars for individual family members are first

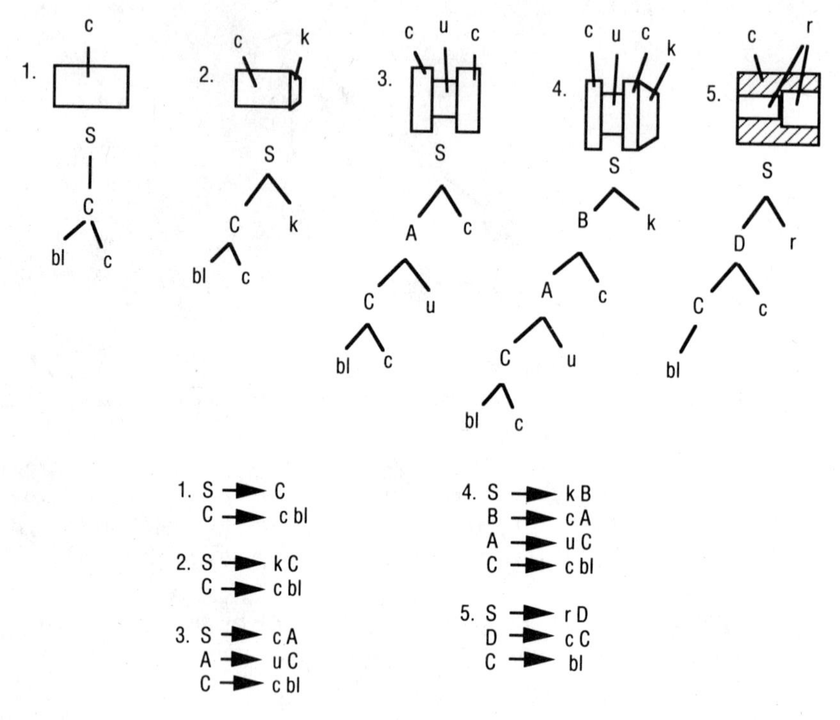

Figure 3.14 A part family and the grammars for it members

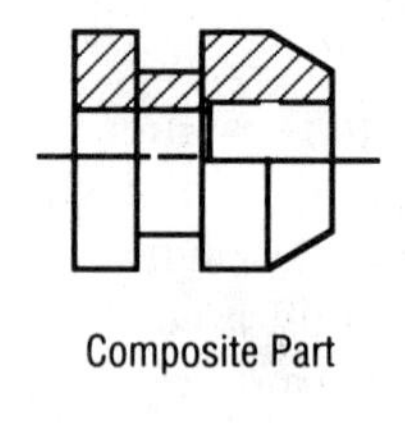

Composite Part

Rule for the composite part:

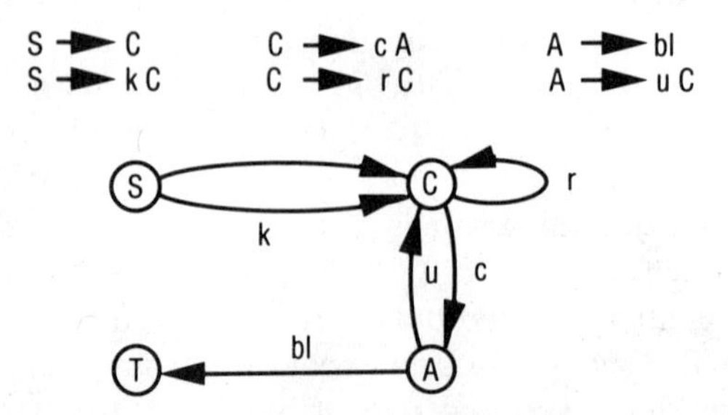

Figure 3.15 Automata for the part family shown in Fig. 3.14

developed. A simplified version of his example is shown in Figures 3.14 and 3.15 to illustrate his approach. In Fig. 3.14, five parts and their corresponding grammar are shown.

The automata developed for the part family is shown in Fig. 3.15. It can be seen as a composite grammar for the entire part family. Any parts falling into this family satisfy the automata.

The syntactic pattern recognition approach is limited by its shape description power. It cannot model complex drawings. This method is limited not only to the classes of features it can recognize, but also by the kind of input it can take. It can only classify an unknown feature but cannot extract it from a drawing. That is to say that it cannot identify a feature from a partial string. Features on a part must be decomposed before they are sent to the recognizer. Actually the string in the first example "BBCCCBBAAABB" cannot be recognized by the grammar given in language: $L(G) = \{ C^nB^mA^n \mid m,n = 1,2,\ldots\}$. The leading and trailing B_s must be removed before the recognition can start.

3.3 The Decomposition Approach

The decomposition approach partitions a design model into several smaller volumes. To be usable the decomposed smaller volumes need to be manufacturing or design features in order that they can be used for process planning. A recognition step is needed after the decomposition step to find the semantics of the features. Unfortunately, so far there is no method which can guarantee the decomposition to produce usable manufacturing or design features. This problem is similar to that of boundary representation in working the constructive solid geometry conversion problem in solid modelling. There is no unique solution for the later problem either [Requicha 1983].

As was discussed in Section 3.1, there are two types of decomposition problems—those for numerical control cutter path generation and those for machining features. For cutter path generation, two strategies can be taken. One is to do cut-and-collect, another strategy is to use macros for features. Since the second strategy uses the results from the machining feature decomposition, further discussion is not necessary. Following, the cut-and-collect strategy is discussed.

Taking a workpiece W and the solid model of a part S, one can find the material to be removed as: $E = W - S$. $E = E_i$, $i=1,2, \ldots n$. One can cut E into small cells (spatial enumeration representation) as shown in Fig. 3.16. The size of each cell is related to the cutter diameter and the depth of cut. That is, the height of the cell is equal to the depth of cut and the width. The length of the cell is equal to the diameter of the tool. Starting from any location, adjacent cells are collective. One may define a direction for cell collection. The direction

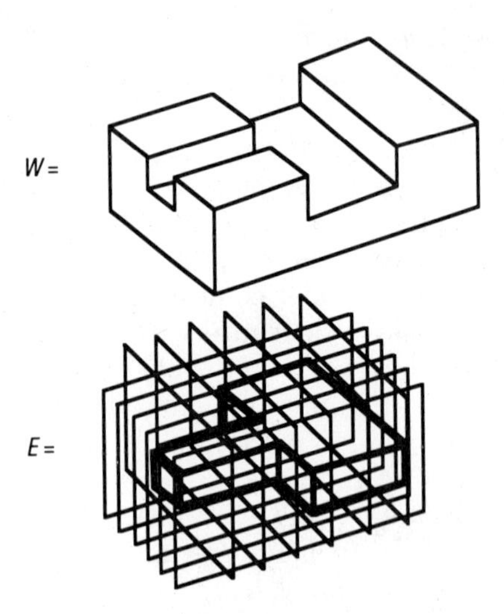

Figure 3.16 Numerical Control cutter path generation using decomposition approach

is the cutter path. When there are disjointed cells, the cutter is either moved up to avoid the solid area or turned to another direction to continue on the current. Some other strategies can also be used to make the process more effective. This approach has been used by the TIPS [Okino and Kubo 1970] system and Armstrong [Armstrong, et al. 1984]. For more details on the decomposition techniques see Tilove [1980].

A second type of decomposition is to decompose E into meaningful E_is, not just spatial cells. Woo [1982] discussed an approach that uses a convex hull (the minimum convex object containing the original object, for more discussion see [1986]) and difference methods to generate an alternating sum of volumes (feature primitives). No implementation was attempted. The idea of this approach is briefly discussed in this section. The procedure begins with the solid model of a part, P. The convex hull of the part, C, is assumed to be the raw stock (not a perfect assumption). The difference between the convex hull and the part is called E. E may contain several disjointed solids. For the next iteration, P is replaced by E. The process repeats itself until either (1) E is the null set, or (2) a cycle is detected. The final result can be written as an alternating sum of volumes:

$$P = C_1 - E_1$$
$$E_1 = C_2 - E_2$$
$$E_2 = C_3 - E_3$$
$$\ldots$$

Therefore, $P = C_1 - (C_2 - (C_3 - (C_4 - \ldots)))$

$$= C_1 - C_2 + C_3 - C_4 + \ldots$$

In the above equation there are both plus and minus signs. The plus sign can be interpreted as a joining (assembly) operation and the minus sign as a machining operation. For machining process planning, we would like to see only minus signs in the equation. By rearranging the equation, the following series can be obtained.

$$P = C_1 - V_1 - V_2$$

where

$$V_i = C_j - C_{j+1} \qquad j = 2 \times i$$

An example of applying this procedure is shown in Fig. 3.17 . In the equation only the difference operator is used. The result can be seen as a Destructive Solid Geometry (DSG) model. Each V_i is a volume to be removed by a machining process. Unfortunately, there

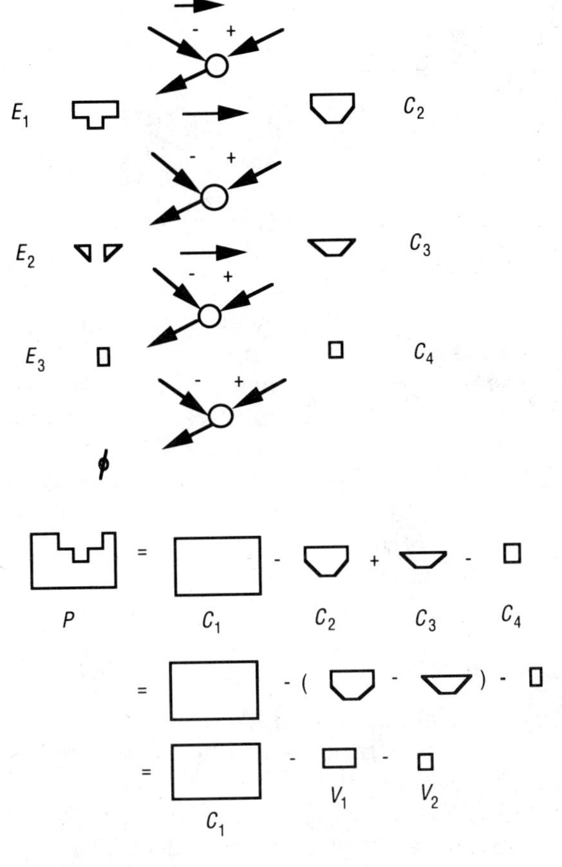

Figure 3.17 Decomposing into machining volumes

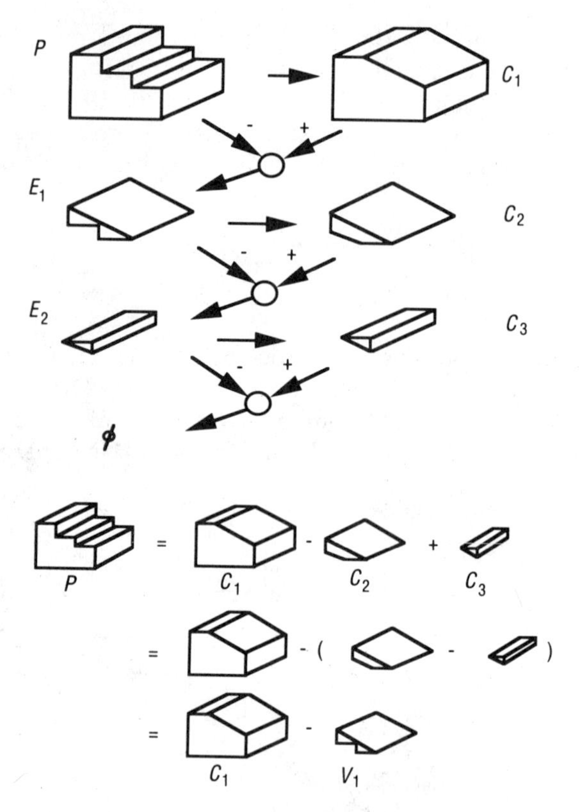

Figure 3.18 Part containing protrusions

is no way to guarantee that either C_1 or V_i are easily obtainable. The raw stock's shape is not selected from an existing set, but is generated by a convex hull operation. When the part consists of only depressions, the raw stock C_1 and volumes V_i are obtainable. However, when the part contains protrusions, unrecognizable features may result (Fig. 3.18). Additional procedures have to be taken to convert a raw stock into C_1. This procedure makes the actual process plan unrealistic. For process planning, a separate feature recognition operation needs to be done on all V_is.

3.4 The Logic Approach

When boundary representation is used as the design representation, the topological structure of features can be defined by logic rules for feature recognition and extraction. This approach was first proposed by Henderson [1984, Henderson and Anderson 1984]. A production rule defining a feature can be written as follow:

IF C_1, C_2 , C_3 , . . . , C_n THEN A

where C_1, C_2 , C_3 , . . . , C_n are conditions, and A is action. If all of the conditions are satisfied, the action is taken. Feature rules can be written to define each feature. A rule includes both the topological and geometrical information. For example, a rule for a simple cylindrical hole can be written in an English form as follows [Henderson 1984]:

IF a hole entrance exists

and the face adjacent to the entrance is cylindrical, and the face is convex,

and the next adjacent face is a plane,

and this plane is adjacent only to the cylinder,

then the entrance face, cylindrical face and plane comprise a cylindrical hole.

The same rule can be translated into predicate calculus Horn clauses. The following rule is in Prolog language format.

```
cylindrical-hole (Set-of-faces) :-
    entrance (Face1)
    adjacent (Face1, Face2),
    cylindrical (Face2),
    convex (Face2),
    adjacent (Face2, Face3),
    not-equal (Face3, Face1),
    plane (Face3),
    adjacent-faces (Face3, Face1),
    Set-of-faces = [Face1, Face2, Face3].
```

Entrance, adjacent, cylindrical, convex, etc., need to be defined as separate rules as well. Most of these relations can be easily found in the boundary model. The rule implementation does not need extensive programming. For example, a rule "entrance" can be defined as a face with an inner loop, and "adjacent" true when there are common edge between the two faces.

The part to be recognized is represented in predicate form. Each element in the boundary representation is represented by a predicate. For example, face, face_loop, edge, vertex, etc., are all predicates. Since there is a one to one mapping between the predicate representation and the boundary representation, the translation from the boundary representation to the predicate representation is easy. For each rule, the Prolog system exhaustively searches through all facts to prove the rule is true. For example, if the facts data base consists of three face facts:

face(f1,plane, ...)

face(f2,cylinder, ...)

face(f3,plane,...)

Then the rule for the simple hole is being activated. First Prolog will try to bond f1 to Face1 in entrance. Let us assume f1 does not contain an inner loop. Entrance(f1) has failed. f2 is tried and also fails. Finally, f3 is tried. Entrance(f3) is true, and f3 is bonded to Face1. Prolog may now proceed to adjacent(Face1,Face2). If Face1 has bonded to f3, Prolog will try to bond f1 to Face2. The process continues until either one of the right hand side rules fails, or the cylindrical-hole rule returns with [f3, ., .] as the set-of-faces.

One can see that the implementation of rules is quite simple, yet it is quite inefficient to do the search, especially when the part is complex. This approach is computational expansive. Although it is easy to develop rules for independent features, it is difficult to develop rules for intersecting features. Henderson [1984] implemented only sweep volume features.

3.5 The Graph-Based Approach

Joshi [1987, Joshi and Chang 1987] developed a graph-based approach for feature recognition. In this approach the boundary representation of the part is transformed into an Attributed Adjacency Graph (AAG) (when represented inside a computer, a graph can be represented by a triangular matrix). An AAG is a graph with attributes assigned to each of the arcs. An AAG can be defined as G = (N,A,T) where N

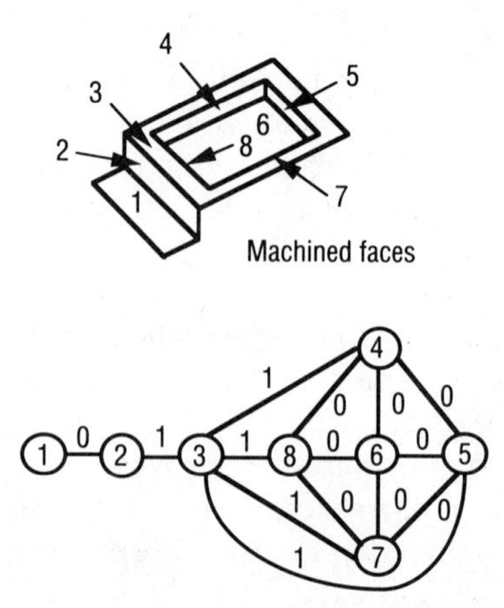

Figure 3.19 Example part and its AAG representation

is the set of nodes, A is the set of arcs, and T is the set of attributes assigned to arcs in A. Each face on the part is represented as a node in an AAG. The adjacency of the two faces is shown as an arc. Wherever two adjacent faces form a convex angle, the arc is labeled 1. Concave adjacency is labeled 0. Figure 3.19 illustrates the AAG of an example part. Part features can also be represented using AAG (Fig. 3.20). The recognition problem is then a graph isomorphism problem which is classified as an NP problem. There is no polynomial time algorithm for the problem, when the problem size (number of nodes) grows the computation time growth will explode.

A simple rule for a step can be written as follows:

IF

graph is linear, and

has exactly two nodes with a incident arc with attribute '0'

THEN

feature is STEP

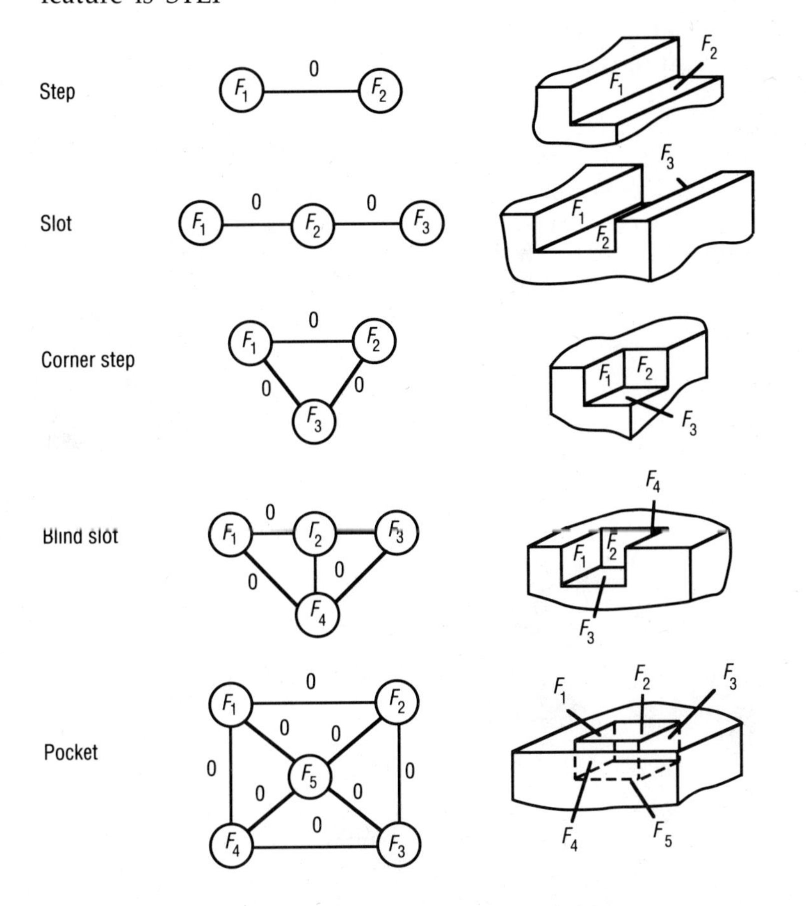

Figure 3.20 Features represented by AAG

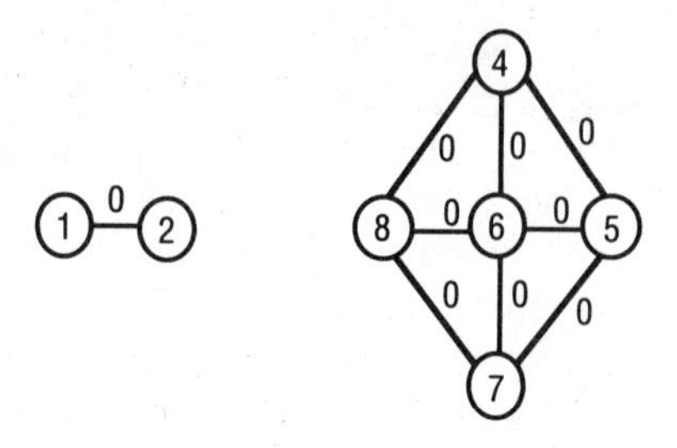

Figure 3.21 Deleting arcs

A rule can be developed for each feature type. Before the search for feature subgraphs begins, a node deleting the procedure is applied. The procedure deletes nodes which only have "1" incident arcs.

This procedure reduces a graph into several disjoined subgraphs. Figure 3.21 shows the results of the procedure after it is applied to the example part in Fig. 3.19. Each subgraph corresponds to a single feature or a set of intersecting features.

Single features can be recognized through a matching procedure. Intersecting features may share some faces; therefore, more processing is required before they can be recognized. Node splitting procedures are used to split two intersecting features. For example, in Fig. 3.22 there are two intersecting slots. Face 1 is shared by both slots, and Faces 2 and 3 were originally the same face. The node splitting procedure will split face (node) 1 into two faces. A node joining procedure will join face 2 and 3 into face 2/3.

This approach is quite effective for 2½ D features. The arc deleting strategy eliminates a lot of unnecessary searches. The ability to split and join nodes allows this approach to recognize intersecting features. However, since this approach does not use the geometry information, a detailed classification of the feature requires another

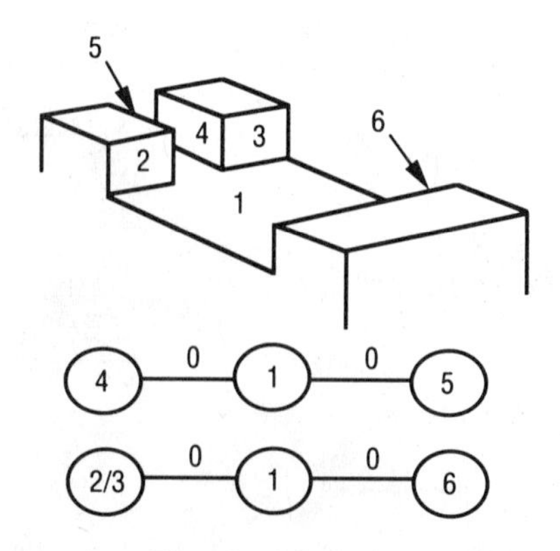

Figure 3.22 Node splitting and joining

step. For example, slots can be recognized and extracted from a part. However, depending on the bottom angles of a slot, it can be a dovetail slot or a simple slot. Additional rules are used to do secondary classification.

3.6 Conclusions

Several different approaches have been used for design/process planning interface. They can produce reasonable results under a very limited domain. The domain is limited to sweep volumes which include 2½ D features. Most of the approaches can recognize only separated features. When a feature is intake as part of an object, these methods may not be efficient or may not even be able to function. More research is needed to solve this general problem.

All of the existing approaches use the syntactic information of the part design for recognition. Geometry and topology are used to define features. In the case of the syntactic pattern approach, the geometry is transformed and represented by a set of pattern primitives. The topology of the part is implied in the production rules. In the logic approach, the feature rules use the geometry and topology as they are defined in a boundary representation. Any set of faces which satisfies the definition is collected as the named feature. The graph-based approach uses a graph to represent the topological information of a part or a feature. For machining process planning purposes, the features defined by the syntactic information seem to be appropriate. Basically, the manufacturing features are those producible by an existing machining process. It would be inappropriate for syntactic information to be used for design analysis since the semantics of each feature also play an important role.

Feature recognition is difficult not only because there is no general theory of geometric reasoning, but also because the definitions of features are imprecise. Although a few features, such as a hole, counterbore, countersink, straight slot, dovetail slot, etc., are generally understood, the definitions of features normally are vague. The same geometric feature may be conceived as several different manufacturing features depending on the direction from which they are viewed and the manufacturing processes available. Therefore, a good feature recognition system must provide users with a means to model the features they are using. As well, a few generic features should be built in, and the rest defined as needed.

In this chapter, several existing ideas on feature recognition and decomposition have been discussed. It is clear that the problem of feature recognition has not been solved. As it is discussed in Chapter 1, a process planning system should not rely on feature recognition alone. A better design modelling system should provide partial feature information. Such information can be used to supplement the

Table 3 Summary of Feature Recognition Studies

Year	Authors	Representation Scheme: Input	Representation Scheme: CAD System Chosen & Internal Representation Scheme	Features	Algorithm/Heuristic
1976	Grayer	B-REP	BUILD	* 2½ D Pockets * NC Tool Path	* divide the pocket into horizontal laminae of constant cross section
1975/ 1976	Woo	CSG	* CSG-tree * Spatial relationships of the primitive volume faces	* Cavity * NC Tool Path	* special language for defining domain features in terms of the primitives and operators used to create the feature
1980	Kyprianou	B-REP	* Syntactic pattern recognition (SPR) * Feature grammar/expression (FG/E)	* SPR ⇒ Protrusions and Depressions * FG/E ⇒ classify shape and generate GT code	* special developed algorithm based on syntactic pattern recognition and feature grammar/expression
1982/ 1983??	Jared	B-REP	* BUILD-4 * define structural primitives and primary faces	* Protrusions and Depressions	* feature grammars (S,T,P,I) * using features to construct object * build simple shape features in the user interface to the BUILD-4 modeler
1982/ 1984	Choi	B-REP	* STOPP * 3D boundary file ⇒ 2½ D description * 2D cross section of round holes of arbitruary profile	* Holes implemented * Pockets, slots, steps, 2D contour, plane, sculptured surfaces, etc. ONLY conceptually classified and proposed	* syntactic pattern recognition * machined surfaces ⇒ pattern of faces
1982/ 1983	Woo	CSG		* Volumes of removed material * NOT form features * limited to polyhedral objects	* convex hull algorithm ⇒ removed material * alternating sum volume ⇒ volume decomposition * recursive but NOT robust

1983	Staley	B-REP	* 2D cross section	* holes' cross section ⇒ mapping to a process plan	* formalized syntactic pattern recognition * strings of a language ⟺ sequences of 2D derected line segment
1982	Jakubowski	B-REP	* special developed part description language	* part shapes	* syntactic pattern recognition ⇒ part description language
1984	Armstrong	B-REP	* 3D cell element	* NC tool path * full 3D parts * Optimal fixturing orientation	* NO features recognized * subdivide part into a sec of 3D cell elements
1984/ 1986	Srinivasan	B-REP	* ROMULUS * context-free grammar $G(V_n V, P, S)$	* limited to pre-defined machined features * selection of machining centers and tools * generative process planning * recognize machining process rather than machined features	* syntactic pattern recognition ⇒ syntax analysis procedure * two B-REP matching * B-REP graph ⇒ tree data structure
1984	Kung		* PADL-I * limited to blocks and cylinders	* disjointed from feature (defined by CAM-I) * generate a general (not practical) process plan	* rule-based expert system * can NOT solve feature intersection problem
1984/ 1985	Henderson	B-REP	* ROMULUS	* cavity features: holes, pockets, slots * exhaustive search strategy, combinatorial problem * GT code generation from recognized features	* logic programming * rule-based expert system implemented in PROLOG

(Continued)

Table 3 Summary of Feature Recognition Studies (Continued)

Year	Authors	Representation Scheme		Features	Algorithm/Heuristic
		Input	CAD System Chosen & Internal Representation Scheme		
1985/ 1987	Lee	CSG		* fillets, rounds, chamfers, necks, bores	* principal axes concept $\Rightarrow$ feature extraction * feature created by adding/subtracting cylindrical primitive solids * unifying manufacturing features $\Rightarrow$ node-pairing * reduce the nonuniqueness in CSG model
1987	Joshi	B-REP	* ROMULUS * AAG	* polyhedral features * cylindrical holes * tool approach direction * feature precedencies	* graph-based approach subgraph (AAG) * concept of machined face and its adjacency * virtual pocket * intersection decomposition

recognition, thus making the recognition task easier. An example of such an approach will be discussed in the last chapter of this book. Finally, a summary table of existing feature recognition studies is shown in Table 3.

References

Aho, A.V., Sethi, R. and Ullman, J.D., *Compilers, Principles, Techniques, and Tools*, Addison-Wesley, Reading, Mass. 1986.

Armstrong, G.T., Carey, G.C., and de Pennington, A., "Numerical code generation from a geometric modeling system," *Solid Modeling by Computers from Theory to Applications*, (M.S. Pickett and J. W. Boyse, eds.), Plenum Press, New York, 1984, pp. 139–157.

Choi, B.K., "CAD/CAM Compatible Tool Oriented Process Planning System," Ph.D. Thesis, School of Industrial Engineering, Purdue University, West Lafayette, Ind., 1982.

Cohen, P.R. and Feigenbaum, E.A., (eds.), *The Handbook of AI*, vol. III, William Kaufmann, Inc. 1982.

Fu, K.S., *Syntactic Methods in Pattern Recognition*, Academic Press, New York, 1974.

Fu, K.S., *Syntactic Pattern Recognition and Applications*, Prentice-Hall, Englewood Cliffs, N.J., 1982.

Grayer, A.R., "The automatic production of machined components starting from a stored geometric description," in *Advances in Computer Aided Manufacturing*, (D. McPherson, ed.), pp. 137–151, North-Holland Publishing Co., Amsterdam, 1977.

Henderson, M.R., and Anderson, D.C., "Computer recognition and extraction of form features: A CAD/CAM link," *Computers in Industry*, vol. 5, 1984, pp. 329–339.

Henderson, M.R., "Extraction of feature information from three dimensional CAD data," Ph.D. Thesis, Purdue University, West Lafayette, Indiana, 1984.

Iwata, K., Kakino, Y., Oba, F., and Sugimura, N., "Development of non-part family type computer aided production planning system CIMS/PRO," *Advanced Manufacturing Technology*, (P. Blake, ed.), pp. 171–184, North-Holland, 1980.

Jared, G.E.M., "Shape features in geometric modeling," *Solid Modeling by Computers from Theory to Applications*, (M.S. Pickett and J. W. Boyse, eds.), Plenum Press, New York, 1984, pp. 121–137.

Joshi, S., and Chang, T.C., "Graph-based heuristics for recognition or machined features from a 3D solid model," CAD, March 1987.

Joshi, S., "CAD interface for automated process planning," Ph.D. Thesis, Purdue Univeristy, 1987.

Kakino, Y. et al., "A new method of parts description for computer-aided production planning," *Advances in Computer-Aided Manufacturing*, North-Holland, 1979.

Kung, H., "An investigation into development of process plans from solid geometric modeling representation," Ph.D. Thesis, Oklahoma State University, 1984.

Kyprianou, L.K., "Shape classification in computer aided design," Ph.D. Thesis, University of Cambridge, Cambridge, U.K., 1980.

Lee, Y.C., and Fu, K.S., "Machine understanding of CSG: extraction and unification of manufacturing features," *IEEE Computer Graphics and Applications*, 1987, Pp. 20–32.

Liu, C.R. , and Srinivasan, R., "Generative process planning using syntactic pattern recognition," *Computers in Mechanical Engineering*, March, 1984, pp. 63–66.

Milacic, V.R., "SAPT—Expert system for manufacturing process planning," *Computer-Aided/Intelligent Process Planning*, (C.R. Liu, T.-C. Chang, and R. Komanduri, eds.), Bound Volume of ASME WAM, Miami Beach, Fl., November 17–22, 1985. PED-vol. 19, ASME, 1985.

Okino, N. and Kubo, H., "Technical information system for computer-aided design, drawing, and manufacturing," Proceedings of the Second PROLAMAT 73, 1970.

Preparata, F.P., and Shamos, M.I., *Computational Geometry, an Introduction*, Springer-Verlag, 1985.

Requicha, A.A.G., "Solid modeling: current status and research directions," *IEEE Computer Graphics and Applications*, vol 3, no. 7, October 1983, pp. 25–37.

Roberts, L., "Machine perception of three-dimensional solids," In J. Tippett (ed.), *Optical and Electro-Optical Information Processing*, MIT Press, Cambridge, Mass., 1965, pp. 159–197.

Rosenfeld, *Picture Languages: Formal Models for Picture Recognition*, Academic Press, New York, 1979.

Staley, S.M., Henderson, M.R., and Anderson, D.C., "Using syntactic pattern recognition to extract feature information from a solid modelling database," *Computers in Mechanical Engineering*, vol 2, no. 2, 1983, pp. 61–65.

Tilove, R.B., "Set membership classification: a unified approach to geometric intersection problems," *IIIE Transactions on Computers*, vol C-29, no. 10, October 1980, pp. 874–883.

Woo, T.C., "Computer understanding of design," Ph.D. Thesis, University of Illinois, Urbana, Ill., 1975.

Woo, T.C., "Feature extraction by volume decomposition," Conference on CAD/CAM Technology in Mechanical Engineering, M.I.T., March 1982; also, Technical Report no. 82-4, Department of Industrial Engineering, University of Michigan, Ann Arbor, Mich., April 1982.

4

Process Knowledge Representation

The function of process planning is to convert a design into manufacturing instructions. In order to do so, the system must possess knowledge about the manufacturing processes. In Chang and Wysk [1985] it is suggested that such knowledge may be represented in several ways, namely, decision tree, decision table, and expert system rules. In this chapter the representations appropriate for expert process planning system construction are discussed. Before we discuss the actual representation of process knowledge, it is worthwhile to evaluate what is process knowledge and how it can be obtained.

Process knowledge is knowledge about the capabilities of a manufacturing process. A manufacturing process cannot operate without using a tool on a machine. The machine and the tool are the means to carry out the process. Therefore, capability is influenced by the tool and the machine. Capability includes both the geometric and technological capabilities. The following list summarizes the most important process capability parameters for process planning.

1. the shapes and size a process can produce
2. the dimensions and geometric tolerances that can be obtained by a process
3. the process constraints both geometric and technological constraints
4. the economics of a process.

For conventional process planning (where human planners are used), all process capability information comes either in the form of experience or in handbook lists and guides. In order to build an expert process planning system we need to collect this information and represent it in an usable way. Before we can collect the information, it is worthwhile to note that process capability is shop specific. Depending on the machines and tools each shop has, the capability of the same process may vary between two shops. Therefore, the information collected from one shop may not be applicable to another shop. Fortunately, the difference is in the data value and not the general form. A general model can be found for all shops, and specific data need to be fitted to this model for each shop. In the first section of this chapter the levels of process knowledge (process capability, from now on we will use the term process knowledge and process capability interchangeable) will be discussed. The division into levels helps us to distinguish the difference between general and specific process knowledge. The second to the sixth sections discuss the actual process knowledge. In section seven examples of process capability rules taken from two different process planning systems are illustrated. Section eight summarizes the chapter.

4.1 Levels of Process Knowledge

There are three levels of process knowledge: a universal level, a shop level, and a machine level. Universal level process knowledge is the knowledge of a process without regard to the individual shop or machines which perform the process. On the universal level, we say that a twist drill can produce a round hole with certain accuracy. This statement is applicable to all twist drill processes and whether they are performed in company A or company B is not considered. Such universal level process knowledge is normally presented in handbooks and textbooks. It represents an aggregate measure of the process knowledge. Beside being educational, it is useful as a starting point for building up knowledge on other levels. In case manufacturing capability is not available in house, and a process planning system is used to evaluate the manufacturability of a design, universal level process knowledge should be used. It is obvious that all the information provided in this chapter is universal level knowledge. Readers should pay extra attention when attempting to use the data for a specific shop.

The second process knowledge level is shop level where additional processing detail is considered. Although one generally can say that a twist drill operation can produce a certain hole diameter with certain accuracy, at the shop level we look at specific machine or improved cutter requirements, which can produce a much smaller hole or obtain a much better accuracy. This knowledge may not be achievable by other companies. Or perhaps certain equipment is older and not as well maintained, thus the knowledge would be less. In either case the shop level knowledge represents the knowledge of the best machine in the shop. Shop level capabilities are those published by companies for internal use. To obtain shop level knowledge one has to collect data on all machines in the shop. Since shop capability changes over time, a periodic review of shop level knowledge needs to be conducted at fixed intervals and after any major equipment change in the shop. For process selection, shop level capability is generally used.

The machine level knowledge is only applicable to a specific machine. Although the shop has certain capabilities, it does not necessary follow that each machine in the shop has the same knowledge. For example, one may find an old milling machine in the same shop with a high precision milling machine. The capabilities are definitely different. Statistical quality control techniques usually are used to obtain the accurate knowledge data. Control charts used in quality control provide good input for the study of tolerance capability. Other capabilities can be obtained by studying the dimension, kinematics, and the tools of the machine. The shop level knowledge is a summary of the machine level knowledge. Machine level knowledge information is important for selecting the specific machine to perform a specific process.

4.2 Fundamentals of Manufacturing Processes

Now let us look at what we mean by process capability. First we will discuss machining process capabilities. Machining processes are used to remove materials from a workpiece. It can be considered as a way to transforming a solid object by removing volume from it (Fig. 4.1). During this transformation, not only volume gets removed, at the same time new surfaces are created. The characteristics of these new surfaces are dependent on the capability of the process applied. A simplistic way of describing this is to model the process capability by the shape (volume) and surface characteristics a process can create.

The volume removed is a subset of the tool sweep volume. Tool

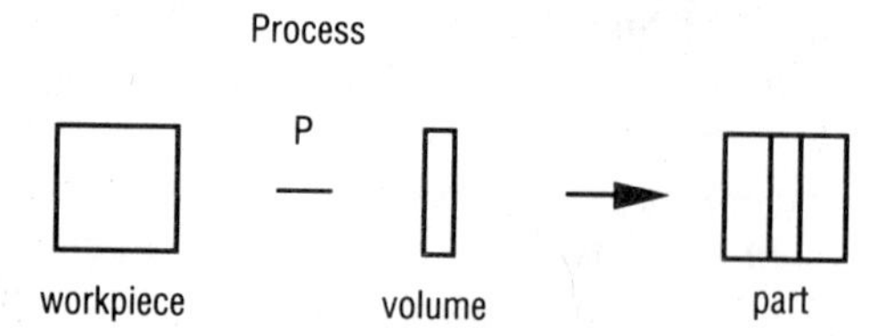

Figure 4.1 Machining process

sweep volume is the volume produced by the tool moving along its cutting path. The intersection between the workpiece and the tool sweep volume is the removed volume. What is left on the workpiece is either a flat surface or a cavity region (a protrusion can be considered as an island in a virtual pocket (Chapter 3)). We can consider the flat surface or the cavity as a machinable feature. This feature consists of several faces (F_m, $m = 1, \ldots, M$). Each face has its shape, dimensions, and a set of surface properties. Therefore, one can consider a machining process as an operator **P** which has the following properties:

$$\mathbf{P}_i(W) \rightarrow W'$$
$$W' := W - Vol_i$$
$$SA' := SA_i$$

where

$\mathbf{P}_i$: Process i
W : Workpiece
W' : Workpiece after machining
Vol_i : Sweep volume created by the process i.
$-$: Boolean difference operator
SA' : Surface property of the newly created feature
SA_i : Surface creating capability of process i.

There are also a set of constraints that need to be satisfied before the process operator can be applied. These constraints can be geometric constraints, such as the accessibility, the relative position of the surface feature, etc. They can also be technological constraints, such as the power consumption, the workpiece deflection, etc. These constraints are imposed by either the tool or the machine. Both the tool and the machine are used to execute the process. For a given process the possible tool design varies only dimension, material, and cutting edge angle; the overall geometry is not changed. For this reason, the geometric constraints usually do not change between tools or machines. However, the technological constraints vary widely between tools or machines.

There are basically two ways a process may produce a shape: *forming* and *generating* [p. 479 Schey 1987]. Both remove materials from the workpiece using one or several cutting edges on the tool. In form machining, the surface created has a negative image of the tool shape. For example a drilled hole takes the negative image of the drill itself. In many other processes, form tools are used for mass production. The accuracy of a surface produced through form machining is dependent on the sharpness of the tool, the tool vibration, the thermal effect during cutting, etc. In the case of generating, the produced shape is created by the sequence of tool sweep motion. For example, a turned surface is usually produced by a single point cutter, and a die cavity is usually generated by a ball end milling cutter with a very complex tool path. In addition errors similar to those that can happen in form machining, the accuracy in generating machining can be affected by path related errors, such as interpolation errors.

Since processes are usually classified loosely, under a general process family there may be several specific processes whose capabilities are quite different. For example, ball end milling can create a very complex surface, yet face milling can only remove a box shaped volume and create a flat surface. Therefore, it is necessary to further classify general processes into more specific processes. The further classification also needs to take into consideration the cutter types used. Table 4.1 shows the classification of some machining processes and the tools and machines used for each process.

4.3 Shape Producing Capabilities

With some machining experience a human being generally can imagine the shape producing capability of a machining process. Although we can't always verbally describe the shapes a process can produce nor an exhaustive list of shapes described by words or sketches, we normally have a good feeling about what can be done. The more experience we have, the better we know what can be done. Exactly how does one determine the process shape producing capability? When we think about it, we may realize that we use our knowledge about the cutting process to reason the shapes producible. It is therefore extremely important for us to understand tool geometry and the mechanism by which a machining process operates.

As discussed earlier, the shape producing capabilities of a process denote the geometrical shape a machining process can produce. A machining process produces a shape with one of the two techniques: form machining and generating machining. In either case, one needs to understand the basic generating process. There are two critical factors in the generating process: a generating surface and a generating motion. The generating surface is the cutting surface created by the cutting edge during tool rotation (self-rotation), and the generat-

Table 4.1 Machining processes, tools, and machines

Process	Sub-Process	Cutters	Machines
	Face milling	Plain	Vertical milling machine
Milling		Inserted-tooth	Horizontal milling machine
	Peripheral milling	Plain	Column-and-knee
		Slitting Saw	Bed type
		Form	Planer type
		Inserted-tooth	Special type
		Staggered-tooth	
		Angle	
		T-slot cutter	Machining center
		Woodruff keyseat cutter	
		Form milling cutter	
	End milling	Plain	
		Shell end	
		Hollow end	
		Ball end	
		Twist drill	Column & upright
		Spade drill	Gang drilling machine
		Deep-hole drill	Radial drilling machine
Drilling		Gun drill	Multispindle drilling machine
		Trepanning cutter	Bench type
		Center drill	Depth hole drilling machine
		Combination drill	
		Countersink	
		Counterbore	
		Shell reamer	Drill press
Reaming		Expansion reamer	Lathe
		Adjustable reamer	
		Taper reamer	
Boring		Adjustable boring bar	Lathe
		Simple boring bar	Boring machine
			Jig bore
	Turning	Plain	Speed lathe
	Facing	Inserted	Engine lathe
	Parting		Toolroom lathe
			Special lathe
Turning	Knurling	Knurling tool	Turret lathe
			Screw machine
	Boring	Boring bars	
	Drilling	Drills	
	Reaming	Reamers	
			Broaching press
			Vertical Pull-down
			Vertical Pull-up
Broaching		Form tool	Surface broaching machine
			Horizontal broaching machine
			Surface broaching machine
			Rotary broaching machine

Table 4.1 Machining processes, tools, and machines (cont'd.)

Process	Sub-Process	Cutters	Machines
Sawing		Hacksaw Bandsaw Circular saw	Reciprocating saw Bandsaw Circular saw
Shaping		Form tool	Shaper Horizontal & Vertical
Planing		Inserted tool	Double housing planer Open-side planer Edge planer Pit Planer
Grinding	Cylindrical grinding Centerless grinding Internal grinding External grinding Surface grinding	Grinding wheels Points	External cylindrical grinder Internal cylindrical grinder Surface grinder Creep feed grinder Tool grinder Disk grinder
Honing		Honing stone	Honing machine
Lapping		Lap	Lapping machine
Tapping		Tap	Drill press Milling machine Machining center

ing motion is the relative motion between the tool and the workpiece. For tools, such as turn tools which do not rotate, the generating surface becomes a generating curve. Figure 4.2 illustrates a form machining generating process.

In the figure, the tool is rather a simple one. It has a curved cutting edge. There is no tool rotation. The generating surface is degenerated to a generating curve. The only tool motion is the in-feed. Since the workpiece rotates around a centerline, the relative motion is

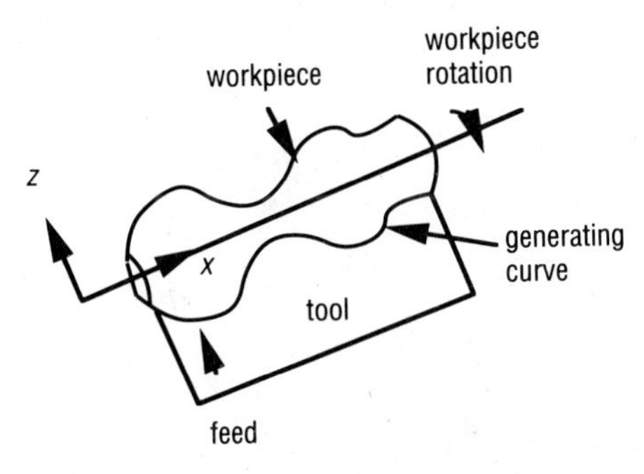

Figure 4.2 Form machining generating process

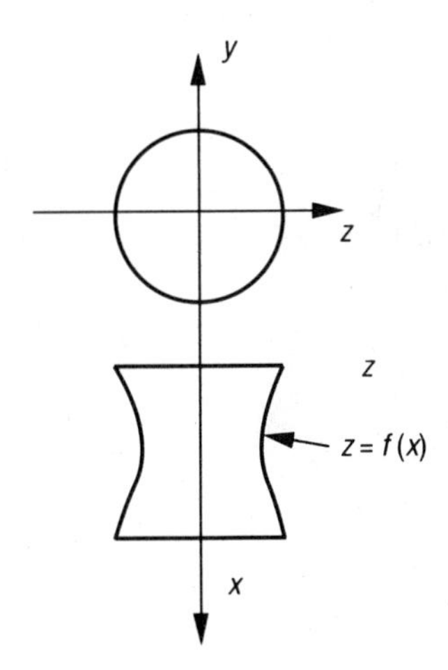

Figure 4.3 Single point cutting, generating curve

a rotation. The generating curve can be described as $z = f(x)$. The shape generated is therefore, $z^2 + y^2 = f^2(x)$.

In the case of generating machining, the process can be more complex. The generating surface can be a point (e.g. single point turning), a curve or a surface. For a single point turning, the tool motion becomes two dimensional, including both in-feed and cross-feed. The cutter path can be described as $z = f(x)$. Taking into consideration the workpiece rotation, the shape generated is the same as the one in the forming cutting.

In milling, boring, reaming, etc., we need to consider the rotation of the tool that produces a generating surface. For example, a face milling tool has a discrete number of inserts (cutting edges); when the spindle starts rotating, these inserts generate a generating surface of the shape of a ring. If the inner and the outer limits of the cutting edge are r_1 and r_0 respectively, and the center of the cutter is located at (x_o, y_o), then the ring can be expressed as: $r_1^2 \leq (x - x_o)^2 + (y -$

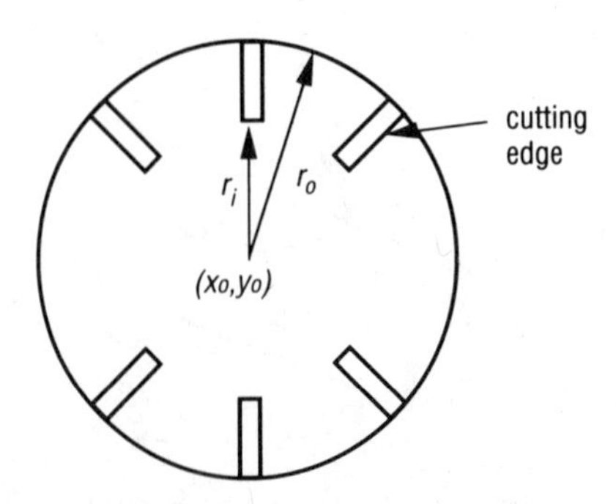

Figure 4.4 Face milling cutter

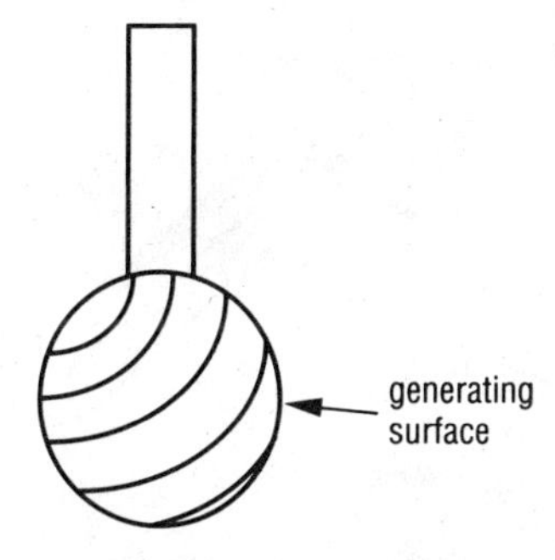

Figure 4.5 Ball end milling cutter

$y_0)^2 \leq r_0^2$ (Fig. 4.4). A ball-end mill has a ball shaped generating surface (Fig. 4.5). However the entire sphere surface cannot be used effectively for cutting. Clearly the top of the sphere where it is joined with the tool shank does not have cutting edge and is not part of the generating surface. The bottom tip of the sphere is also not part of the generating surface. Because the spindle axis passes through the bottom tip, the effective cutting speed at the tip is zero (cutting speed is $2 \pi r n$, where r is the effective radius at the cutting edge, and n is the spindle rpm). When such a tool is moving along a complex path to create a complex shape, there is no simple mathematical description for the produced shape.

The shapes created by those processes are discrete, and each discrete shape element is created by either a linear or a circular sweeping of the generating surface. For example, in Fig. 4.6 a milling cutter executed three linear motions. Although the tool motion is very simple, the sweep volume can no longer be described by a simple mathematical formula. Given the geometry of the tool and the tool path, one may be able to find the sweep volume or the sweep surfaces by using a solid modeller. One can easily see that there are infinite number of final volumes which can be created by those processes. It is not feasible to consider the shape producing capability by combining the generating surface with a complete set of tool paths. Even the forward process capability operator **P** is not easily attainable. The idea of finding a mathematically sound inversed process capability operation is thus not feasible for most cases.

From the discussion above, it is obvious that the shape producing capability of a machining process is determined primarily by the geometry of the cutting tool used and the possible tool motion. Each process has its special way of removing material, therefore, there is only a finite set of tool geometries and tool motion elements one can expect. For example, in Fig. 4.7 a summary of the basic cutting edge arrangement and tool motion for some processes is shown. The darkened edges in the figure denote the cutting edges. The arrow

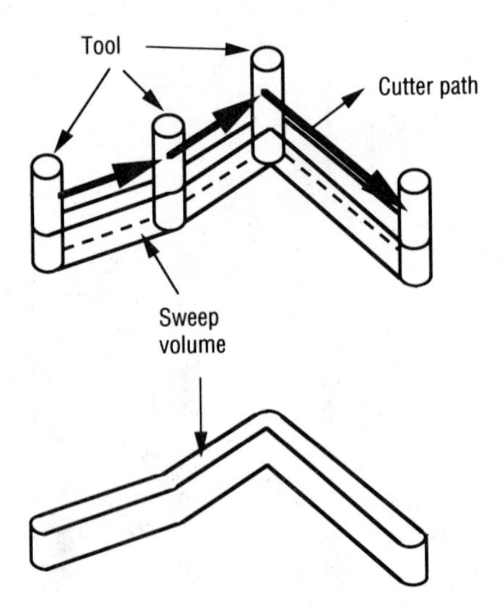

Figure 4.6 Sweep volume created by a milling process

shows either a tool rotation or a tool feed. For example, for drilling there are basically two cutting edges arranged at an angle. The tool rotates about its axis and the cutting motion is a down feed motion. A ball end mill has a much more complex cutting edge arrangement and also allows a wider range of tool motions. From Table 4.1, one can find that each process has a limited set of cutters. By studying the tools and motion, some special characteristics of the shapes producible by a process can be found. Except when special form tools are used, the shapes producible by a process are limited. The feasible cutter path is constrained by the way cutting edges are arranged on a cutter. The cutting edge also determines the generating surface.

In the discussion we presented the produced shape in terms of cutting tools and tool motion. It can be seen as a forward process operation which performs a forward transformation to produce the part. However, for the process planning purpose, what we need is an inversed transformation. For a given shape, we want to determine the feasible processes and tools. There does not seem to be a simple inversed operation. The alternative is to match the processes capabilities representation with the design specifications. Therefore, several other representation methods must be explored.

Since a complete mathematical description has not yet been found useful, most existing process planning systems use symbolic representations. That is to say, the shape producible by a process is assigned a name (symbol). The design is converted into a representation using the same symbols. This conversion is done either manually (most common), or by a CAD interface (see Chapter 3). In this representation, only gross shape description is considered. If they are not

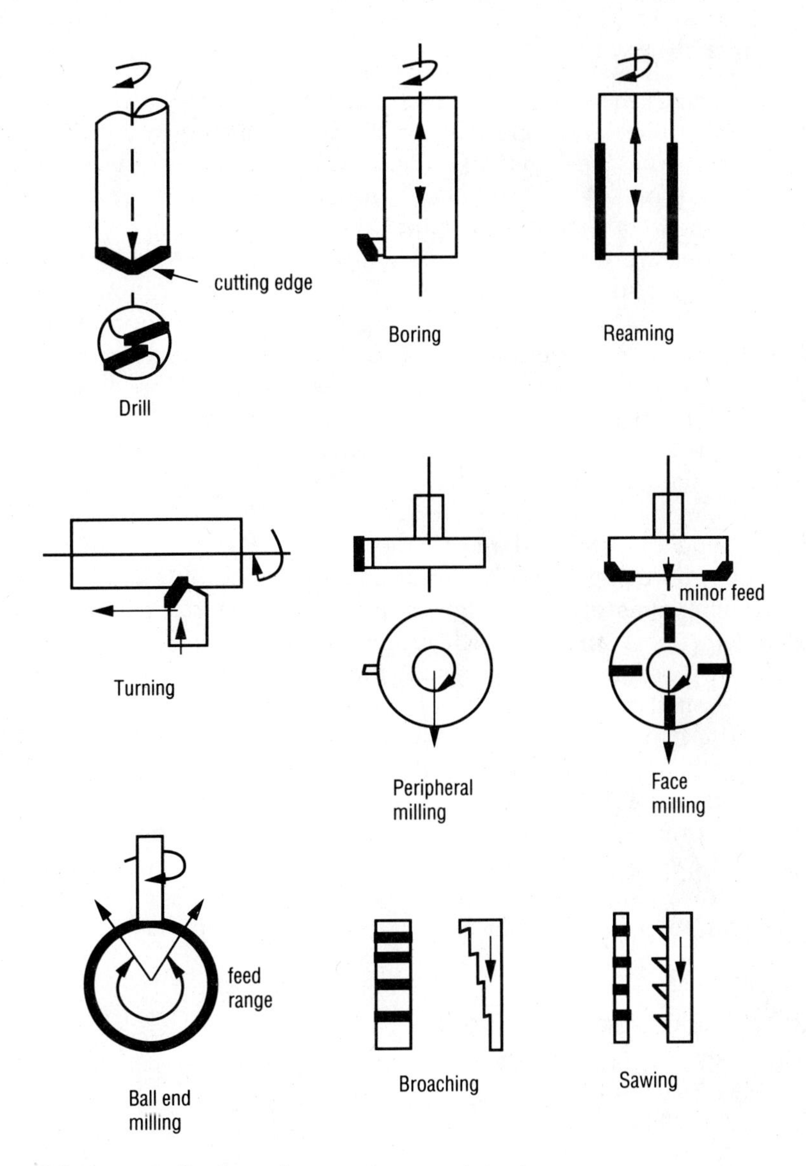

Figure 4.7 Cutting edge geometry and feed

significant to the selection of a process, local details are usually ignored.

Basically, there are three methods for representing the shape producing capability: by edge, by surface, and by volume. All of them can be considered as modelling with features (edge feature, surface feature, and volume feature). More details on feature-based design can be found in Chapter 2.

4.3.1. Representing by Edge

The idea of representing by edge is suitable for two dimensional problems. Turning and hole making are two domains where this representation has been used [Wang et al. 1986, Srinivasan and Liu 1987, Staley et al. 1983, Choi 1982]. Turned part features are very similar to those of a hole except that the former are external features and the latter are internal ones. For turning, there are in-feed and cross-feed—two feed elements that control the shape. Among hole making processes, drilling and reaming use forming cutting, and boring is similar to turning except it only has one controllable feed. Since the shape produced is always symmetric around an axis, a 2D cross section can fully represent the shape.

One can represent the cross sectional boundary by edge primitives (see Figs. 3.3 and 3.5) easily. In the following table (Table 4.2) shape producing capabilities of turning and drilling are shown. Single point turning can create shape elements **B, C, D, ..., P**. Therefore, the shape producing capability of a single point turning is a shape with combinations of those edge primitives. Facing can create only shape elements **A** and **E**, which represent the flat end surface. **A** models the portion of surface about the center, and **E** models the portion of surface below the center. Simple drilling can produce a shape pattern that is represented in a shape language $\mathbf{E}^n\mathbf{D}^m\mathbf{B}^m\mathbf{A}^n$. $\mathbf{E}^n$ and $\mathbf{A}^n$ define the side wall, and $\mathbf{D}^m$ and $\mathbf{B}^m$ model the drill tip. The depth of the hole is defined by n (n edge primitive depth, where each edge primitive has a unit length). The diameter and the tip angle of the hole produced is defined by the variable m. A detailed explanation of shape primitives can be found in Chapter 3. It is worthwhile to point out that although we presented shape capability

Table 4.2 Shape representation with edges

Process	Sub-Process/Tools	Shape Producing Capabilities
Turning	Single point turning	**B, C, D, F, G, H, M, N, O, P**
	Facing	**A, E**
	Parting	**ACE, AGE**
	Knurling	**C, G, B, H, F, D** (w/knurling attribute)
Drilling	Twist drill Spade drill Deep-hole drill Gun drill	$\mathbf{E}^n\mathbf{D}^m\mathbf{B}^m\mathbf{A}^n$
	Center drill	$\mathbf{D}^m\mathbf{B}^m$
	Countersink	$\mathbf{E}^n\mathbf{D}^m\mathbf{E}^o\mathbf{D}^p\mathbf{B}^p\mathbf{A}^o\mathbf{B}^m\mathbf{A}^n$
	Counterbore	$\mathbf{E}^n\mathbf{C}^m\mathbf{E}^o\mathbf{D}^p\mathbf{B}^p\mathbf{A}^o\mathbf{C}^m\mathbf{A}^n$

using shape grammar, in actual implementation, methods other than shape grammar may be used.

It is important to know that there are additional constraints that are not shown in the table. For example, for facing the edge primitives are **A** and **E**. Yet **A** and **E** can be the wall of a hole as well! Therefore, one has to add additional notes to indicate that facing can produce a turned part surface with shape element **A** and **E**.

Although this method may be effective for processes which produce symmetric surfaces or features, it is difficult to use it to describe the complex surface generation capability of many other processes. For example, how can one represent the three dimensional surface producing capability of a milling process with only edges? Even three dimensional shape primitive and grammar may not be a solution. The major advantage of this approach is the simplicity of the representation. However, it is also the major disadvantage of the approach—too simple to model three dimensional features.

4.3.2 Representing by Surface

Since not all shapes generated by a process can be represented by edges, one needs to look beyond edge representation. One way to model the shape producing capability of a process is to analyze the surfaces generated by it. Since the surfaces generated can be extremely complex, it is a easier to consider only feature surfaces. Feature surfaces are characteristic that can be produced by the process being studied. The feature surfaces are created by the generating motion. For example, the corner rounding (Fig. 4.8) of a pocket can be produced only by a cylindrical shaped cutting edge. The tool diameter d_t must be less than or equal to two times the smallest corner radii r_i, $d_t \leq 2\ min(r_i \mid i = 1, \ldots n)$. n: number of corners. The radius of the tool generating surface either matches the corner radii, or a circular interpolation is done to generate the corner. When the bottom of the pocket needs to be rounded, a ball end mill is used. The radius of the ball must meet the designed pocket bottom radii.

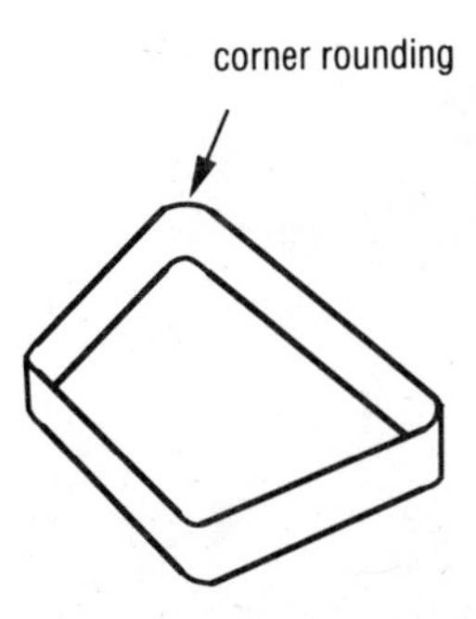

Figure 4.8 Corner rounding

As one can reason that the surface is generated by a generating process, it must take the shape from this generating process. The two major factors—generating surface (cutting edges and tool rotation) and tool motion (in-feed and cross-feed), are dominating factors. From Table 4.3; the surface producible by different processes can be reasoned. Processes like drilling, boring, reaming, broaching, and sawing have only one feed direction; the surfaces generated are either cylindrical (for drilling, boring and reaming), flat or form shape (for sawing and broaching). Turning can have two feed directions, thus the surface is two dimensional. However, this surface must be rotational. Peripheral milling and face milling can move in two dimensions, their motion is restricted to a planar surface, so a flat surface is the major surface type they can produce. Peripheral milling cutters can be formed, so formed slot shapes (e.g., V-slot) can be produced. Since both peripheral milling and face milling tools are circular, they both can produce a round corner. A ball-end milling cutter has a three dimensional feed direction range, it can produce a large variety of surfaces.

The above discussion actually focuses on the local feature of a shape. The global feature is more difficult to model since it involves a set of local features and their connectivity. Srinivasan and Liu [1987] proposed to represent this set of features using a tree grammar. The shapes which a process can produce are modeled by a multi-dimensional tree grammar. This tree grammar for a process model can also be represented in a diagram form. By applying this grammar, many shapes producible can be generated. The grammar for

Table 4.3 Process Surface Producing Capabilities

Process	Sub-Process	Cutters	Surface Capabilities
	Face milling	Plain	flat surface
Milling		Inserted-tooth	
	Peripheral milling	Plain	flat surface
		Slitting Saw	slot surface
		Form	formed surface
		Inserted-tooth	
		Staggered-tooth	
		Angle	
		T-slot cutter	T-slot surface
		Woodruff keyseat cutter	internal groove surface
	End milling	Plain	pocket, slot, flat, round corner
		Shell end	
		Hollow end	sculptured surface, flat, round
		Ball end	corner
Drilling		Twist drill	internal cylindrical surface
		Spade drill	internal cylindrical surface

Table 4.3 Process Surface Producing Capabilities (cont'd.)

Process	Sub-Process	Cutters	Surface Capabilities
		Deep-hole drill	deep internal cylindrical surface
Drilling (Cont'd.)		Gun drill	deep internal cylindrical surface
		Trepanning cutter	large internal cylindrical surface
		Center drill	shallow internal conical surface
		Combination drill	multiple diameter intl cyl surface
		Countersink	internal clipped conical surface
		Counterbore	internal cylindrical surface flat bottom
		Shell reamer	internal cylindrical surface
Reaming		Expansion reamer	internal cylindrical surface
		Adjustable reamer	internal cylindrical surface
		Taper reamer	internal cylindrical surface
Boring		Adjustable boring bar	internal cylindrical surface
		Simple boring bar	internal cylindrical surface
	Turning	Plain	external rotating surface
	Facing	Inserted	end surface
	Parting		two end surface
Turning	Knurling	Knurling tool	knurling surface
	Boring	Boring bars	internal cylindrical surface
	Drilling	Drills	internal cylindrical surface
	Reaming	Reamers	internal cylindrical surface
			flat surface
			open slot surface
			step
Broaching		Form tool	polyhedral through hole surface
			formed surface
			internal keyway
			spline surface
		Hacksaw	flat or curved surface
Sawing		Bandsaw	flat surface, curved surface
		Circular saw	flat surface
		Form tool	flat surface, slot surface
Shaping			irregular hole, rack gear surface
		Inserted tool	flat surface
Planing			V groove
	Cylindrical grindling	Grinding wheels	external cylindrical surface
	Centerless grinding	Points	internal cylindrical surface
Grinding	Internal grinding		flat surface
	External grinding		
	Surface grinding		
Honing		Honing stone	internal cylindrical surface
Lapping		Lap	flat surfaces
Tapping		Tap	threaded surface

a process is also called a process model. This approach is not without difficulties. As discussed earlier, a process may be able to generate a very large number of surface types. Figuring out how to take all of them into consideration in a process model is not easy. Also, to completely represent a boundary representation, a graph is needed. The tree grammar represents a tree that is only part of a graph. Although the geometry may be modeled, only partial topology can be represented. If we try to represent the shape using a graph then we are representing it using volume.

Wu [1988] represented the shape producing capability of a process by the cutter shape (generating surface). For example, end milling cutters are classified into three types: sphere end, torus end, and flat end. Each cutter also has several parameters to characterize its dimensions. A flat end tool is said to be able to produce flat surfaces that do not have adjacent bottom corner radius but have side corner radius. A torus end cutter can produce a flat surface with an adjacent bottom corner radius. The corner radius produced by a torus end cutter is the same as the torus radius. In this work, Wu did not consider the global feature interaction.

From the above examples, you can see that stays on the local representation level are still feasible for many processes. The producible shapes can be characterized by a set of characteristic surfaces. These surfaces should be divided into a AND set ($S_a = \{S_i \mid i = 1,2,\ldots n\}$) and an OR set ($S_o = \{S_i \mid i = 1,2,\ldots m\}$). The OR set contains all surface elements producible by the process; any of them may be produced. The AND set are surface elements which are always produced by the process as a by-product. For example, the surfaces in the AND set for a ball-end mill is a curved surface. For twist drilling, the AND set includes the hole bottom conical surface. Of course, by using this the topology of shape produced is not considered at all. Representing by surface does not burden the system with the difficult problem of exact representation.

4.3.3. Representing by Volume

Most researchers thus far model process shape producing capabilities with machinable volumes [Wysk 1977, Chang 1980, 1982]. Usually these machinable volumes are volumetric features. The difficulty of using volumetric features is how to represent them. The exact representation of a volumetric feature can best be done using a solid model. However, a solid model can be very complex and inflexible to use during process planning. In most generative process planning systems and expert process planning systems, the machinable volume features are coded with a symbolic name, e.g., hole, slot, pocket, etc. Simplicity is the reason for using qualitative representation of volumetric features. However, there is vagueness in this representation due to the imprecise definition of these features. Some precise yet not too complex representation approaches are needed. A few examples can be found in Chapter 3 under the logic approach and the graph approach. With a precise definition of features, the methodology discussed in Chapter 3 can be used to recognize them. Unfortunately, this problem remains unsolved. The other alternative is to use the qualitative approach. When this fuzzy representation is used, a human user needs to interpret the design model for the system. Pictorial examples of feature shapes have to be provided to the users during training as well as during the subsequent operation time in order to eliminate confusion.

Volumetric representation models the global shape producing capability versus that of local shape features for the surface representation. It not only represents the final surface produced by the process but also the entire volume of material removed. When volumetric representation is used in conjunction with a solid modeler, it is possible to take into account the intermediate workpiece shape.

From Table 4.4 you see it is difficult to represent the shape producing capability of some processes with volumetric features. For example, the volume removed by a turning process is rather complex. However, the process capability can easily be modeled with the edge (a generating curve). Yet for drilling, reaming, and boring the volumetric representation is clearly simpler. The question of which representation to adopt in an automated process planning system can only be answered with the consideration of the part domain and the process domain.

Table 4.4 Volume Producing Capabilities

Process	Sub-Process	Cutters	Volume Capabilities
	Face milling	Plain	flat bottom volume
Milling		Inserted-tooth	
	Peripheral milling	Plain	flat bottom volume
		Slitting Saw	slot
		Form	formed volume
		Inserted-tooth	
		Staggered-tooth	
		Angle	
		T-slot cutter	T-slot
		Woodruff keyseat cutter	Internal groove
	End milling	Plain	pocket, slot, flat
		Shell end	
		Hollow end	
		Ball end	sculptured surface, flat
		Twist drill	round hole
		Spade drill	round hole
		Deep-hole drill	deep round hole
Drilling		Gun drill	deep round hole

Table 4.4 Volume Producing Capabilities (cont'd.)

Process	Sub-Process	Cutters	Volume Capabilities
Drilling (Cont'd.)		Trepanning cutter	large round hole
		Center drill	shallow round hole
		Combination drill	multiple diameter round hole
		Countersink	countersink hole
		Counterbore	counterbore hole
Reaming		Shell reamer	thin wall of round hole
		Expansion reamer	thin wall of round hole
		Adjustable reamer	thin wall of round hole
		Taper reamer	thin wall of round hole
Boring		Adjustable boring bar	thin wall of round hole
		Simple boring bar	thin wall of round hole
Turning	Turning	Plain	?
	Facing	Inserted	disk
	Parting		disk
	Knurling	Knurling tool	?
	Boring	Boring bars	think wall of round hole
	Drilling	Drills	round hole
	Reaming	Reamers	thin wall of round hole
Broaching		Form tool	flat bottom volume
			slot
			step
			polyhedral through hole
			formed through volume
Sawing		Hacksaw	?
		Bandsaw	
		Circular saw	
Shaping		Form tool	flat bottom volume, slot
Planing		Inserted tool	flat bottom volume
Grinding	Cylindrical grindling	Grinding wheels	?
	Centerless grinding	Points	internal cylindrical of round hole
	Internal grinding		flat bottom volume
	External grinding		
	Surface grinding		
Honing		Honing stone	?
Lapping		Lap	most surfaces
Tapping		Tap	threaded wall of hole

4.4 Dimension, Tolerance, and Surface Properties Capabilities

Other than the shape producing capability, each process/tool also has it own dimension, tolerance, and surface properties producing capabilities. It is obvious that the drilling process can not drill an infinitely large and deep hole. Neither can it drill a hole infinitely small in size. As a matter of fact, it can only produce holes with discrete size increments. Beside the size capability, every process also has tolerance and surface finish producing capabilities. For process planning purposes we would like to collect this information and represent it in a way which we can use it effectively. In this section we will discuss the capabilities of some processes.

First, we will discuss the dimension capability. The dimension capability is determined by both the tool size and/or the machine tool work envelope. For a process which uses a form generating method, the dimension capability usually is determined by the tool dimension. For example, the hole diameter and the depth producible by a drilling process are determined by the available drill sizes. For generating machining, the dimension is not only limited by the tool but also by the machine tool where the process is conducted. On three axis machining, usually the Z axis is the spindle axis. If a cavity is being machined, the maximum depth is limited by the tool length. When trying to go any deeper, the spindle will begin to interfere with the workpiece. However, if the cavity opening is big enough to allow the spindle to go in, then the limitation is the maximum travel of the spindle. For the X and the Y axis, the dimension limits are the machine travel limits. It is worth noting that there are three levels of process capability as discussed in Section 4.1. Each shop has its own limit as what can be produced. Shop A may be able to drill a hole of 0.01″ diameter, but shop B can do 0.1″ only. A database of available tools and machines must be kept current in order to supply such information to the process planning system.

For hole producing processes, the depth of hole machinable is also related to the diameter of the tool used. An ordinary drilling process can drill a depth from three to eight times the hole diameter. However, in practice, the limit of the depth/diameter ratio is four. Any hole that has a depth/diameter greater than this has to be drilled by a deep hole drilling process such as a gun drill. This capability is limited by the deflection of the tool, the friction between the cutter and the hole wall, and the chip flow.

The cause of tolerance capability is more complex. Many factors affect the accuracy of a process, i.e. tool wear, tool deflection and chatter, thermal deformation of machine tool elements, tools and workpiece, control inaccuracy, round out of tool assembly, fixture error, etc. The tolerance capability is caused by a combination of these factors. It is not possible to predict the tolerance precisely. Therefore the only feasible way is to rely on the experience base. From various handbooks and textbooks, tolerance data can be collected. For a specific shop the capability has to be modified by collecting data within the shop. Table 4.5 shows some tolerance and surface finish capability information summarized from several sources

Table 4.5 Tolerance, surface finish, etc., capabilities

Process	Sub-Process	Cutters	Tolerances, surface finish, etc., capabilities		
				roughting	finishing
Milling	Face milling	Plain	tol	0.002	0.001
		Inserted-tooth	flatness	0.001	0.001
	Peripheral milling	Plain	angularity	0.001	0.001
		Slitting Saw	parallelism	0.001	0.001
		Form	surface finish	50	30
		Inserted-tooth			
		Staggered-tooth		roughting	finishing
		Angle	tol	0.002	0.001
		T-slot cutter	flatness	0.001	0.001
		Woodruff keyseat cutter	surface finish	50	30
		Form milling cutter			
	End milling	Plain		roughting	finishing
		Shell end	tol	0.004	0.004
		Hollow end	parallelism	0.0015	0.0015
		Ball end	surface finish	60	50

Process	Sub-Process	Cutters	Tolerances, surface finish, etc., capabilities					
Drilling		Twist drill Spade drill Trepanning cutter Center drill Combination drill Countersink Counterbore	length/dia = 3 usual = 8 maximum mtl < Rc 30 usual mtl < Rc 50 maximum					
			Dia	Tolerance			usual	best
			0 - 1/8	+0.003	−0.001	True position	0.008	0.0004
			1/8-1/4	+0.004	−0.001	roundness	0.004	
			1/4-1/2	+0.006	−0.001	surface finish	100	
			1/2-1	+0.008	−0.002			
			1-2	+0.010	−0.003			
			2-4	+0.012	−0.004			
		Deep-hole drill Gun drill	Dia	Tolerance		surface finish > 100		
			<5/8	0.0015		straightness 0.005 in 6 inch		
			>5/8	0.002				

Reaming		Shell reamer Expansion reamer Adjustable reamer Taper reamer	Dia / Tolerance: 0-1/2 0.0005 to 0.001 1/2-1 0.001 1-2 0.002 2-4 0.003 (roughting / finishing) roundness 0.0005 / 0.0005 true position 0.01 / 0.01 surface finish 125 / 50
Boring		Adjustable boring bar Simple boring bar	length/dia 5 to 8 Dia / Tolerance (roughing / finishing): 0-3/4 0.001 / 0.0002 3/4-1 0.0015 / 0.0002 1 - 2 0.002 / 0.0004 2 - 4 0.003 / 0.0008 4 - 6 0.004 / 0.001 6 -12 0.005 / 0.002 straightness 0.0002 roundness 0.0003 true position 0.0001 surface finish 8
Turning	Turning Facing Parting Knurling Boring Drilling Reaming	Plain Inserted Knurling tool Boring bars Drills Reamers	diameter tolerance to 1.0 0.001 1 - 2 0.002 2 - 4 0.003 surface finish 250 to 16
Broaching		Form tool	tolerance 0.001 surface finish 125 to 32
Sawing		Hacksaw Bandsaw Circular saw	length tol / squareness / surface finish / cutting rate / material: 0.01 / 0.2 / 200-300 / 3-6 sq in/min / to Rc45 0.01 / 0.2 / 200-300 / 4-30 sq in/min / to Rc45 0.008 / 0.2 / 125 / 7-36 sq in/min / to Rc45

Table 4.5 Tolerance, surface finish, etc., capabilities (cont'd.)

Process	Sub-Process	Cutters	Tolerances, surface finish, etc., capabilities
Shaping		Form tool		roughting | finishing location tol 0.005 | 0.001 flatness 0.001 | 0.0005 surface finish 60 | 32 (cast iron) surface finish 125 | 32 (steel)
Planing		Inserted tool	(as above)
Grinding	Internal grinding Cylindrical grinding Centerless grinding External grinding Surface grinding	Internal	Dia | Tolerance: roughing | finishing 0 - 1 | 0.00015 | 0.00005 1 - 2 | 0.0002 | 0.00005 2 - 4 | 0.0003 | 0.0001 4 - 8 | 0.0005 | 0.00013 8 - 16 | 0.0008 | 0.0002
		center ground and centerless	roughing | finishing tolerance 0.0005 | 0.0001 parallelism 0.0005 | 0.0002 roundness 0.0005 | 0.0001 surface fin 8 | 2
		flat	roughing | finishing tolerance 0.001 | 0.0001 parallelism 0.001 | 0.0001 surface fin 32 | 2

Honing		Honing stone	Dia / Tolerance: roughing / finishing 1: +0.0005-0.0 / +0.0001-0.0 2: +0.0008-0.0 / +0.0005-0.0 4: +0.0010-0.0 / +0.0008-0.0 surface finish 4 roundness 0.0005
Lapping		Lap	roughing / finishing tolerance 0.000025 / 0.000015 flatness 0.000025 / 0.000012 surface fin 4-6 / 1-4
Tapping		Tap	tolerance 0.003 roundness 0.003 surface fin 75

[Bralla 1986, DeGarmo, et al. 1984, Drozda and Wick 1983, Metcut 1980]. For example, for a normal hole drilling process (including twist drill and spade drill), the diameter tolerance for a drill of 1/8 inch diameter is +0.004 −0.001. When the drill diameter is 2 inches, the diameter tolerance becomes +0.012 −0.004. The greater the tool diameter the worse the diameter tolerance. Also, for the drilling process, the true position tolerance is usually greater than 0.008″. These data can be represented in a rule combined with other process capability variables. Examples can be found in Section 4.7.

4.5 Process Constraints

When modeling the capability of a machining process, it is not sufficient to consider only the shape and dimension capabilities. A process must satisfy some other constraints before it can be applied. These constraints are prerequisites for the process to be effective. There are basically two types of constraints: geometric constraints and technological constraints.

4.5.1 Geometric Constraints

Geometric constraints are those constraints which can be identified by the geometric relations of features. There are several causes of these geometric constraints.

a. Interference between the non-cutting part of the tool with the workpiece.

b. Technological reason.

The tool geometry-caused limitations are those due to the non-cutting part of the tool. Many of these constraints are caused by the interference between the non-cutting part of the tool with the workpiece. For example, one can not bore, ream, nor broach a hole before a hole is already in existence. From Fig. 4.7, it is clear that boring and reaming tools that have a cutting edge on the peripheral of the tool can remove only a thin layer of the hole wall. The end of the tool will interfere with the workpiece surface if no hole has been drilled already. When milling a cavity, if there is a recess which is deeper than the tool reach, the tool will not be able to cut it either. The shank of the tool or the spindle will touch the workpiece before the cutting edge is able to reach the recess. Global part design information is needed in order to ensure that there is no interference between the tool and another part of the geometry during cutting.

Some constraints are actually caused by technological reasons. Because they are all geometrically related, they are considered under geometric constraints. One of these constraints is drilling a hole on a slant surface. Due to the slippage of the drill bit, the position and the shape of the hole are difficult to control. Another example is to drill a longitudinal hole on a thin and long strip. The cutting force and torque will twist and deform the strip. Therefore, where is the hole located and how is it related to the adjacent surfaces is critical to the process.

Some geometric constraints for machining processes are listed below and also shown in Fig. 4.9. This list is by no means complete. It can be considered as a small sample of what actually exists.

1. You cannot drill on a slant surface due to the drill bit slippage problem. Several things can be done to prevent the problem. If the slant surface is also machined, one can drill the hole before milling the slant surface. If the slant surface already exists, a milled counter bore may be machined.

2. You cannot drill a flat bottom hole with twist drill due to the drill bit's conical shape constraint. The cutting edge of a drill is arranged at an angle. The bottom of the hole produced must also carry the same angle. When a flat bottom is desired, an end mill can be used to flatten the bottom.

3. You cannot tape to the bottom of a flat bottom hole (must leave some clearance) due to tap structure. The tip of a tap is not effective to reach the flat bottom.

4. You cannot bore nor ream before a hole is already in existence due to tool structure. The bottom of the boring bar and reamer is the non-cutting part of the tool. Tool cannot plunge into the material without a center hole.

5. You cannot drill two holes too close to each other due to uneven cutting force, slippage and potential wall damage may occur.

6. You cannot mill a pocket before a hole is plunged due to the tool structure. A hole is drilled first, or a hole is slowly milled with an end mill. Milling cutters need side clearance to remove chips. When breaking a pocket, there is no opening for chip removal.

7. You cannot drill when the location of the hole is too close to the wall. The thin wall between the hole and the side of the work-piece may break during drilling.

8. You cannot use certain tools for recess because they cannot reach.

9. You cannot broach a blind hole nor a blind slot due to tool structure limitation. A broach is pulled from both sides. The cutting edge on a broach is on part of one face of the tool. It can machine either an open surface or enlarge an existing hole (usually not round hole).

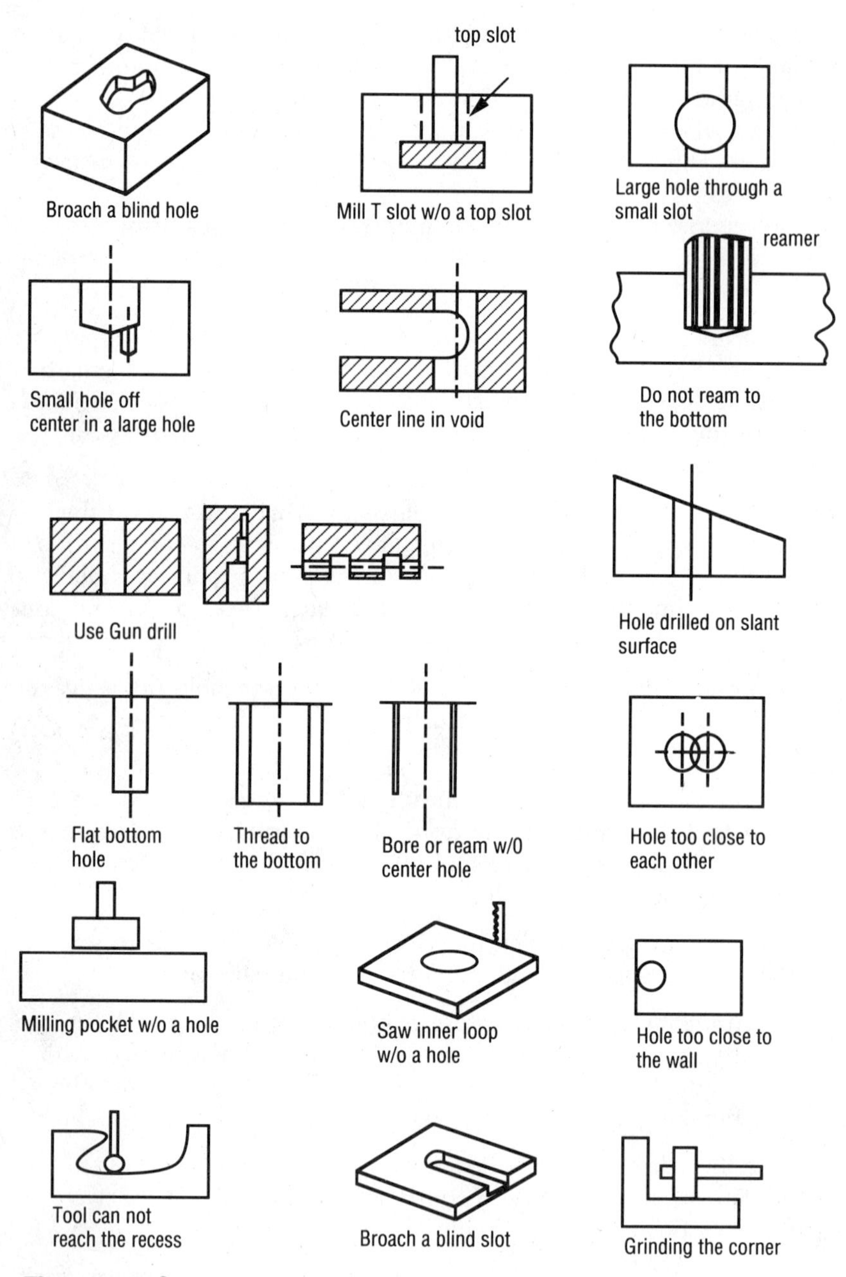

Figure 4.9 Geometric constraints

10. You cannot saw an enclosed loop unless a hole is drilled first due to tool structure limitation for the same reason as a broach.
11. You cannot mill a T-slot before a top slot is milled due to tool structure limitation. To mill the horizontal bar of the T, there must be a vertical slot for the tool shank clearance.

Although we can expand this list, few of the existing process planning systems consider them. In order to take these constraints into consideration, a sound geometric reasoning scheme must be developed. The process planning system must repeatedly consult the geometric reasoning module for detailed constraint information. Without considering these data, the generated process plan can be erroneous. It is clear that to conduct geometric reasoning for this purpose, a solid model is almost unavoidable. Algorithms need to be developed to detect these geometric constraints. One example system is the QTC [Chang, et al. 1988] which will be discussed in detail in Chapter 6.

4.5.2 Technological Constraints

Technological constraints are those constraints due to power consumption (HP), cutting force (F_c), etc. There are also constraints related to the deflection of workpiece and the strength of materials which are not considered here because they are not process related (fixturing related). These constraints can be used to determine the fixturing method and the machine tool needed for carrying out the operation. The cutting force affects the stability of the fixturing. The fixture must hold the workpiece against the cutting force. Cutting force is also directly related to the power consumption. Each machine can deliver only a certain amount of power. These data need to be known before a process plan can be completed.

Many variables contribute to the cutting force. For a given tool, the feed (f), speed (v), depth of cut (a_f & a_p), width of cut (b_w), cutting fluid, workpiece material, and the sharpness of the tool, all affect the cutting force requirement. To predict the cutting force precisely in a production environment is not possible. Fortunately for the planning purpose, one does not really need a precise estimation of the cutting force value. A rough estimate can work well in this case. The rough estimation can be done using an empirical formula shown in Table 4.6. Parameters used in the formulae are obtained through experiment. They are determined by the sharpness of the tool, the tool geometry, and tool and workpiece materials. After the cutting force has been determined, the power consumption can be calculated by the cutting force times cutting speed divided by 33,000 times the machine efficiency (η_m).

Another method of estimating power is to use the material re-

Table 4.6 Cutting force, and power consumption

Process	Sub-Process	Cutting Force F_c (lb)	Power (hp)
Milling	Face milling	$K_F\ \nu^{\alpha F} a_f^{\beta F} a_p^{\gamma F}\ D_t^{\delta F} b_w^{\epsilon F} z^{\xi F}$	$\frac{F_c\ V_c}{33{,}000\ \eta_m}$
	Peripheral milling End milling	$K_F\ f^{\beta F}\ a_p^{\gamma F}\ D_t^{\delta F}\ b_w\ z$	
Drilling		$K_F f^{\beta F}\ a_p^{\gamma F}\ D_t^{\delta F}$ torque	$\frac{T_s\ rpm}{63{,}030\ \eta_m}$
Reaming			
Boring		$K_F\ f^{\beta F}\ a_p^{\gamma F}$	$\frac{F_c\ V_c}{33{,}000\ \eta_m}$
Turning	Turning		
	Facing		
Shaping			
Planing			
Broaching		$K_F\ a_p^{\beta F}\ D_m z_c$	$\frac{F_c\ V_c}{33{,}000\ \eta_m}$

where

V_c: cutting speed fpm
η_m: machine efficiency
T_s: Torque

Table 4.7 Material removal rate

Process	Sub-Process	MRR	
Milling	Face milling Peripheral milling End milling	$W\ a_p\ f\ n\ N$	*W*: width of the cutter *ap*: depth of cut *f*: number *n*: number of teeth *N*: spindle rpm
Drilling		$(\pi\ D^2/4)\ f\ N$	*D*: tool diameter
Boring		$12\ V\ f\ a_p$	
Turning	Turning	$12\ V\ f\ a_p$	
	Facing	$6\ V\ f\ a_p$	
Broaching		$12\ tr\ W\ V\ n$	*tr*: rise per tooth *W*: Width of the tool *V*: cutting speed *n*: number of tooth in contact with part
Shaping		$L\ t\ f\ N_s$	*L*: strock length N_s: number of stock per minute

Table 4.8 Determining power requirements in machining

Material	Hardness	Unit Power* hp/in³/min: Turning P_t HSS and Carbide Tools (feed. 005-.020 ipr)		Drilling P_d HSS Drills (feed .002-.008 ipr)		Milling P_m HSS and Carbide Tools (feed .005-.012 ipt)	
	Bhn	Sharp Tool	Dull Tool	Sharp Tool	Dull Tool	Sharp Tool	Dull Tool
Steels, Wrought and Cast	85-200	1.1	1.4	1.0	1.3	1.1	1.4
Plain Carbon	35-40 R_c	1.4	1.7	1.4	1.7	1.5	1.9
Alloy Steels	40-50 R_c	1.5	1.9	1.7	2.1	1.8	2.2
Tool Steels	50-55 R_c	2.0	2.5	2.1	2.6	2.1	2.6
	55-58 R_c	3.4	4.2	2.6	3.2†	2.6	3.2
Cast Irons	110-190	0.7	0.9	1.0	1.2	0.6	0.8
Gray, Ductile and Malleable	190-320	1.4	1.7	1.6	2.0	1.1	1.4
Stainless Steels, Wrought and Cast	135-275	1.3	1.6	1.1	1.4	1.4	1.7
Ferritic, Austenitic and Martensitic	30-45 R_c	1.4	1.7	1.2	1.5	1.5	1.9
Precipitation Hardening Stainless Steels	150-450	1.4	1.7	1.2	1.5	1.5	1.9
Titanium	250-375	1.2	1.5	1.1	1.4	1.1	1.4
High Temperature Alloys	200-360	2.5	3.1	2.0	2.5	2.0	2.5
Nickel and Cobalt Base							
Iron Base	180-320	1.6	2.0	1.2	1.5	1.6	2.0
Refractory Alloys							
Tungsten	321	2.8	3.5	2.6	3.3†	2.9	3.6
Molybdenum	229	2.0	2.5	1.6	2.0	1.6	2.0
Columbium	217	1.7	2.1	1.4	1.7	1.5	1.9
Tantalum	210	2.8	3.5	2.1	2.6	2.0	2.5
Nickel Alloys	80-360	2.0	2.5	1.8	2.2	1.9	2.4
Aluminum Alloys	30-150 500 kg	0.25	0.3	0.16	0.2	0.25	0.4

Table 4.8 Determining power requirements in machining (cont'd.)

Material	Hardness	Unit Power* hp/in³/min					
		Turning $P_†$ HSS and Carbide Tools (feed.005-.020 ipr)		Drilling P_d HSS Drills (feed .002-.008 ipr)		Milling P_m HSS and Carbide Tools (feed .005-.012 ipt)	
	Bhn	Sharp Tool	Dull Tool	Sharp Tool	Dull Tool	Sharp Tool	Dull Tool
Magnesium Alloys	40-90 500 kg	0.16	0.2	0.16	0.2	0.16	0.2
Copper	80 R_B	1.0	1.2	0.9	1.1	1.0	1.2
Copper Alloys	10-80 R_B	0.64	0.8	0.48	0.6	0.64	0.8
	80-100 R_B	1.0	1.2	0.8	1.0	1.0	1.2

*Power requirements at spindle drive motor, corrected for 80% spindle drive efficiency.

†Carbide

moval rate (MRR) and the unit power (P), $HP = MRR \times P$. The material removal rate can be calculated from the cutting parameters. A formula for the material removal rate can be found in Table 4.7. The average unit power requirements are determined by workpiece and tool material type, hardness, feed, and sharpness of the tool. Table 4.8 shows the average unit power requirements for turning, drilling, and milling. These data can be used as a starting point only. By no means are they precise.

Example:
A hole is being drilled by a new HSS drill of 1″ diameter. The workpiece material is low carbon steel 1118 with BHN hardness 120 R_c. The recommended feed is 0.022 ipr and the speed is 105 fpm. The machine efficiency is 0.8. What is the power required?

$$n = \frac{105 \times 12}{1 \times \pi} = 401 \text{ rpm}$$

$$V_f = f\,V = 0.022 \times 401 = 8.82 \text{ imp}$$

$$MRR = (\pi D^2/4)\ V_j$$

$$= (\pi\ 1^2/4)\ 8.822 = 6.93\ in^3/min$$

From Table 4.8 the unit power $P = 1.0$.

$$HP = 1.0 \times 6.93 = 6.93 \text{ hp}$$

4.6 Process Economics

Another extremely important factor is process economics. It is always of interest to us to find the most economical solution. Often, it means the survival of the company. Basically, process economics means the cost efficiency of the processes. For mass production, it is necessary to go through a very detailed economic analysis before selecting a specific processing method. However, for the usual small to medium batch production it is not practical to conduct a very detailed study. The amount of time spent cannot justify the saving. Some rough estimation or just common sense should be used to select the best process. Whenever there are more than two candidate processes that are both technologically capable for the task, it is time to compare their relative costs. A process cost model can be stated as:

$$C = \text{Labor cost} + \text{Machine overhead} + \text{Tool change cost} + \text{Tool cost}$$

$$C = C_m(T_m + T_h) + (C_t + C_m\,T_t)\,\frac{T_m}{T_l}$$

where

C : total cost for the operation ($)
C_m: operator rate and machine overhead ($/hr)
C_t: cost of tool ($)
T_m: processing time (hr)
T_h: material handling time if any (hr)
T_t: tool change time (hr)
T_l: tool life (hr)

The equation, T_m / T_l indicates the number of tool changes for the operation. It is determined by the tool life and the processing time. Processing time can be calculated by the necessary tool travel divided by the feed speed. For example, for drilling a x inch deep hole using a feed speed of a ipm, $T_m = x / a / 60$.

The tool life equation can be expressed as (for milling):

$$T_l = \frac{C}{V^{\alpha} f^{\beta} a_p \gamma}$$

where:

C: a constant determined by the tool geometry, tool material and workpiece material.
V: the cutting speed, fpm.
f: the feed, ipr.
a_p: the depth of cut, inch.
α, β, γ: coefficients.

Unfortunately, several difficulties prohibit us from using this model to predict the operation cost. First, the cutter path is not known at the process selection time. It will be very time consuming, if it is even possible, to generate the cutter path for all candidate processes for each feature to be machined. The second problem is the availability of coefficients for each pair of tool material/ workpiece material. There are very few published data for tool life equations. Most of the tool life / machinablity data are published in terms of recommended feed and speed. With these two major problems, one can conclude that this approach will not work for real life problems. One must find a quick-and-dirty way to estimate the cost.

Since we are dealing with the machining of a single feature, it is reasonable to assume that the material handling time is negligible. The chance of changing a tool during the operation is also minimal. Also, usually the feed and speed recommended by the Machining Data Handbook [Metcut 1980] make the tool life about 60 minutes. Since the recommended feed and speed are what are used in most machining operations, it is reasonable to assume that $T_l = 1\ hr$. Therefore the cost function can be simplified to:

$$C = (C_m + C_t)\ T_m\ /\ 60$$

The machining time can be estimated by the material removal rate (MRR) and the volume (V) to be removed (current feature volume).

$$T_m = Vol\ /\ MRR \cdot 60$$

Therefore, the cost function can be re-written as:

$$C = (C_m + C_t)\ Vol\ /\ MRR \cdot 60$$

The maximum material removal rate data can be estimated using the equations given in Table 4.7. Before it can be estimated one still needs to find the tool size, feed, and speed. The volume to be removed can be estimated much easier than the length of the cutter path. Cost data can also be obtained from the accounting department. With these data the processing cost can be calculated. This cost information can be used to select the most economical process to use for machining a volumetric feature. Since in the processing cost function two variables—C_t and MRR are related to the process, other capabilities of interest are tool cost and material removal rate. These capabilities should be used for selecting economical machining processes.

Example:

Continue the example in Section 4.5.2. The hole to be drilled is 3″ deep. The machine and operator rate is \$40/hr. The tool cost is \$10 each. What is the production cost of the hole?

$$Vol = \frac{\pi\ 1^2}{4}\ 3 = 2.356\ \text{in}^3$$

$$C = (40 + 10)\ \frac{2.356}{6.93 \cdot 60} = \$0.283$$

The above model does not consider the fixed cost of tooling. The tool cost used in the model is the incremental tool cost. In case special tools are needed, a fixed cost may occur. In that case, the fixed cost must be evenly distributed to the entire batch of parts made.

4.7 Examples of Process Capabilities Representation

So far we have discussed the representation of process capabilities. In this section the knowledge base taken from two different process planning systems are presented. The examples show how process capability representations can be implemented. They don't, by any means, represent the best way to model process capabilities. As a matter of fact they don't even include all the capability data discussed in the previous sections. The first example is taken from a system called TIPPS [Chang 1982, Chang and Wysk 1985]. The second example is taken from the process planning module AMPS [Kanumury 1988] of an integrated system—QTC [Chang, et al. 1988]. We would like to point out that capability data used in the examples do not necessary match exactly with those summarized in the Table 4.4. What in Table 4.4 are "generic" and in the examples they are "specific." We will briefly explain the process capability representation in each system.

4.7.1 The TIPPS Process Capability Rules

The TIPPS system process selection module contains a knowledge base and an interpreter. The input to the system is a solid model with recognized surfaces. Each set of recognized surfaces are part of a 3D solid feature. The identifier of each surface set is assigned by a human user. Process knowledge is contained in the process knowledge base and modeled using a language called PKI. PKI is a rule language which allows conditions to be written in a post-fix format. A compiler compiles the rule base into a format that is later used by the interpreter. Each process is described by a rule which has a **IF...THEN...** format.

Two rules are presented in the following. The first rule is a rule for twist drilling. The code (numerical code) for the process is "1". Feature "hole" is assigned a code "111". The symbols used in the rule construction are: "!" means retrieve from the design data base. "+", "−", "*", "/", "<", ">", "<=" and ">=" are regular mathematical operators. Since the system uses stack for operation, some stack operators are also provided. "@" is a store operator. "DUP" is an operator which duplicate the top item on the stack. The rule can be read as: **IF** the shape of the current object is a hole, the length of the object ≤ 12 * the diameter, etc. **THEN,** the selected process is "1" (twist drilling), the tool diameter is the current hole diameter, and the job is done. For the rough end mill operation, the shape can be flat (201), step (202), straight slot (203), pocket (207), or curved surface (209). We also have a few tolerance and surface constraints. The last clause in both rules is (free) which breaks the loop.

```
;
; twist drilling (code 1)
; 111: hole
( ( if
   (shape ! 111 = )
   ( length ! 12.0 diameter ! * <= )
   ( diameter! 0.0625 >= )
   ( diameter! 2.000 <= )
   ( tlp ! diameter ! 0.5 ** 0.007 * >= )
```

```
    ( tln ! diameter ! 0.5 ** 0.007 * 0.003 + >= )
    ( straightness ! length ! diameter ! / 3. ** 0.0005 * 0.002 + >= )
    ( roundness ! 0.004 >= )
    ( parallelism ! length ! diameter ! / 3. ** 0.001 * 0.003 + >= )
    ( true ! 0.008 >= )
    ( sf ! 100 >= )
  )
  ( then
    ( 1 process @ )
    ( diameter ! dtl @ )
    ( free )
  )
  )
  ;
  ;
  ; rough end mill ( code 13 )
  ;
  ( ( if
    ( shape ! dup 201 = dup 202 = dup 203 = dup 207 = 209 = )
    ( tlp ! .004 >= )
    ( tln ! .004 >= )
    ( parallelism ! .0015 >= )
    ( sf ! 60 >= )
  )
  ( then
    ( 13 process @ )
    ( free )
  )
  )
```

4.7.2 The AMP/QTC Process Capability Rules

The AMP system uses an expert system shell—KEE® (Knowledge Engineering Environment) as its implementation tool. There are two parts to each rule, the constraints and the action. Actually what is shown in the examples are LISP functions used in the rules. For example, for the drilling process, the function drilling_satisfy_constraints is the condition portion of the rule. The drilling_modify_constraints and drilling_modify_radius are the action portion of the rule. In the rule, "feat" is a variable representing the volumetric feature being evaluated. "get.value" is a macro which retrieves the slot value from an object frame. "get.values feat 'thru_type 'own world" means get the value of "thru_type slot from "feat" frame. From the example, the drilling process is able to produce "conical thru" type feature. After the process is selected, the constraints are relaxed. The operation is very similar to that in TIPPS. The second rule is for rough face milling. The shape producible is "FLAT".

```
;
; process knowledge for drilling
;
(defun drilling_satisfy_constraints (feat)
  (let ((world *GLOBAL.WORLD*))
(if (and
  (in.range.radius feat world 0.055 3.5)
  (member (car (get.values feat 'thru_type 'own world)) '(CONICAL THRU))
  (>= (get.value feat 'surface_finish 'own world) 125)
  (>= (get.value feat 'radial_tol 'own world) 0.008)
  (>= (get.value feat 'locational_tol 'own world) 0.008)
  (>= (get.value feat 'cylindricity_tol 'own world) 0.01)
  (>= (get.value feat 'circularity_tol 'own world) 0.01)
  (>= (get.value feat 'parallelity_tol 'own world) 0.01)
  (>= (get.value feat 'perpendicularity_tol 'own world) 0.01)
  (>= (get.value feat 'angularity_tol 'own world) 0.01)
  (>= (get.value feat 'concentricity_tol 'own world) 0.01)
  (<= (get.value 'work_piece 'hardness 'own world) 369)
  (<= (get.value feat 'l_d_ratio 'own world) 6.0))
t
nil)))
(defun drilling_modify_constraints (feat world)
  (put.value feat 'surface_finish 500 'own world)
  (put.value feat 'radial_tol 0.1 'own world)
  (put.value feat 'locational_tol 0.1 'own world)
  (put.value feat 'circularity_tol 0.1 'own world)
  (put.value feat 'cylindricity_tol 0.1 'own world)
  (put.value feat 'parallelity_tol 0.1 'own world)
  (put.value feat 'perpendicularity_tol 0.1 'own world)
  (put.value feat 'angularity_tol 0.1 'own world)
  (setq *GLOBAL.WORLD* world))
(defun drilling_modify_radius (ucval lcval feat world)
  (modify.radius '(3.5 0.055) (list ucval lcval) feat world))
;
; Process Knowledge for Face Milling
;
(defun rough_facemilling_satisfy_constraints (feat)
(let ((world *GLOBAL.WORLD*))
(if (and
  (member (car (get.values feat 'b_type 'own world)) '(FLAT))
  (>= (get.value feat 'surface_finish 'own world) 125)
  (>= (get.value feat 'locational_tol 'own world) 0.005)
  (>= (get.value feat 'parallelity_tol 'own world) 0.004)
  (>= (get.value feat 'perpendicularity_tol 'own world) 0.004)
  (>= (get.value feat 'angularity_tol 'own world) 0.004)
```

```
    (<= (get.value 'work_piece 'hardness 'own world) 369)
    (>= (get.value feat 'w_d_ratio 'own world) 0.5))
  t
  nil)))

(defun rough_facemilling_modify_constraints (feat world)
  (put.value feat 'surface_finish 500 'own world)
  (put.value feat 'locational_tol 0.1 'own world)
  (put.value feat 'parallelity_tol 0.1 'own world)
  (put.value feat 'perpendicularity_tol 0.1 'own world)
  (put.value feat 'angularity_tol 0.1 'own world)
  (setq *GLOBAL.WORLD* world))
```

4.8 Conclusions

In this chapter process knowledge representation has been discussed. The capability of a machining process is described by the shape, dimension, tolerance, and surface properties it produces, its geometric and technological constraints, and the economics of the process. Several data tables which contain universal level process knowledge were presented. Finally, two examples taken from systems the author developed were used to illustrate how such process knowledge can be implemented in a real process planning system. From the discussion it can be seen that shape capability is the most difficult to represent in an automatic process planning system. In order to fully utilize the shape knowledge, either a customized design system or an automatic interface needs to be implemented. Since it is very difficult to model the shape producing capability of a machining process with simple mathematical formulae, most of the existing systems used symbolic description of the shapes.

The dimension, tolerance, and surface properties capabilities are easier to represent. These capabilities are data value estimates which can be represented by a data value. However, when looked at closely, one may find that the amount of data needed are numerous. There is a permutation of these capability values based on the tool type, tool material, workpiece material, tool dimension, and machine tool used. What makes it worse is the fact that every shop has it own capability which is everchanging. This is part of the reason why it is difficult to build a generative process planning system for a larger domain of processes and parts types.

The next capability variable of interest is the geometric constraints. Geometric constraints, like shape capability need to deal with the geometric representation. The constraints should be described in a way that can be used by the process planning system to check actual geometric relations on the design model. This is again very difficult due to difficulty in geometric reasoning. The majority of existing systems either do not consider geometric constraints or consider only a few special cases.

The last capability variable is the cost of the process. To be accurate, the cost calculation needs to include labor cost, machine overhead, tool change cost, and tool cost. Since some of the variables stay constant for the same machining feature, the machining time, tool life, tool change time, and tool cost need to be considered when comparing two candidate processes. Since the final cost analysis requires information not only on the process, but also on the tools, tool path, and machine, it can not be done during the process selection stage. Most of the existing systems use a relative cost factor for process selection. Processes are ranked based on the relative cost (preference). The relative cost is assigned based of the experience on the economics of a process.

In order to have a system intelligent enough to make correct decisions about process selection, the process capability knowledge discussed in this chapter must be represented in a proper way. Although most of the existing systems do not use complete information, readers should try to consider including more knowledge in the systems to be built. Some useful data for building a process capability knowledge base are included in the Tables. Readers are encouraged to build a small system with these data. Additional data can be found from [Schey 1987, DeGarmo, et al. 1984, Bralla 1986, Metcut 1980, and Drozda and Wick 1983].

References

Boothroyd, G., *Fundamentals of Metal Machining and Machine Tools*, McGraw-Hill, New York, 1975.

Bralla, J.G. (ed), *Handbook of Product Design for Manufacturing*, McGraw Hill, New York, 1986.

Chang, T.C. and Wysk, R.A., *An Introduction to Automated Process Planning Systems*, Prentice-Hall, Englewood Cliffs, N.J., 1985.

Chang, T.C., Anderson, D.C., and Mitchell, O.R., "QTC—An integrated design/manufacturing/vision inspection system for prismatic part," 1988 Computers in Engineering Conference, ASME, San Francisco, Calif., July 31–Aug. 3, 1988.

Choi, B.K., "CAD/CAM compatible tool oriented process planning for machining centers," Ph.D. Thesis, Purdue University, December 1982.

DeGarmo, E.P., Black, J.T., Kohser, R.A., *Materials and Processes in Manufacturing*, 6th ed., Macmillan, New York, 1984.

Drozda, T.J. and Wick, C., *Tool and Manufacturing Engineers Handbook, Vol 1 Machining*, Society of Manufacturing Engineers, 1983.

Kanumury, M., Shah, J., and Chang, T.C., "An automatic process planning system for a quick turnaround cell—An integrated CAD and CAM System," USA-Japan Symposium on Flexible Automation, ASME, July 18th, 1988.

Metcut, *Machining Data Handbook, 3rd Ed., Vol 1 and 2,* Machinability Data Center, Metcut Research Associates Inc., Cincinnati, Ohio, 1980.

Schey, J.A., Introduction to Manufacturing Processes, 2nd ed., McGraw-Hill, New York, 1987.

Srinivasan, R., and Liu, C.R., "On some important geometric issues in generative process planning," *Intelligent and Integrated Manufacturing Analysis and Synthesis,* ed. C.R. Liu, A. Requicha, and S. Chandrasekar, Bound Volume, ASME Winter Annual Meeting, Boston, Massachusetts, December 13-18, 1987, pp 229–243.

Staley, S.M., Henderson, M.R., and Anderson, D.C., "Using syntactic pattern recognition to extract feature information from a solid modelling database," *Computers in Mechanical Engineering,* vol. 2, no. 2, 1983, pp. 61–65.

Wang, C.-L., Bagchi, A., and Ahluwalia, R.A., "DMAP: A computer integrated system for design & manufacture of axisymmetric parts," Bound Volume, Knowledge-Based Expert Systems for Manufacturing, ASME, Winter Annual Meeting, Anaheim, Calif., December 7–12, 1986, pp. 327–338.

Wu, M.C., "A New Methodology for Automatic Process Planning and Execution Based on Adaptive Information Modeling," Ph.D. Thesis, Purdue University, W. Lafayette, Ind., 1988.

5

Expert System Formulation

So far we have discussed the design representation of a mechanical part, the CAD interface methods, and the actual knowledge needed for doing process planning. Now we need to see how to put all this information together into a system. There are as many ways one may design a process planning system as there are ways one may design a mechanical part. As mentioned in Chapter 1, in general we classify process planning approaches into variant and generative approaches. A variant approach utilizes Group Technology as the means for part representation and process plan retrieval. Process planning knowledge is not coded explicitly into the system. In case of a generative approach, several methods have been used to model the process knowledge and the plan generative concept. There are those systems which use a decision tree, decision table, or expert system as the tool for modelling the process knowledge. These systems may use several different problem solving strategies, i.e., feature-based approach using either composite component concept or generic process capability concept, surface-based approach, etc. In this chapter we will attempt to focus on how to build an expert process planning system. Since process planning is a knowledge intensive activity, it seems most appropriate to take the expert system approach.

Before any detail is discussed, we first need to see what an expert system is. An expert system is *"A computer program using expert knowledge to attain high levels of performance in a narrow problem area."* [Waterman 1986] The above definition is rather vague. In order to clarify the definition we must state some characteristics which an expert system should have.

a. It separates knowledge itself from how to use the knowledge. In traditional computer programming, the knowledge and the control are mixed in the code.

b. The knowledge is codified symbolically, and the system is able to reason symbolically. Traditional programming converts everything into numbers. Algorithms work on numbers to come up with an answer.

c. It can explain its reasoning process. There is no way to know how a traditional program comes up a specific answer. Although one may be able to trace all the steps a program goes through, the program cannot tell why it goes through those steps.

d. It must perform like a human expert. Even software that satisfies all the above three conditions can only be called a knowledge-based system, and not an expert system, if it cannot perform up to the level of a human expert.

Generally, an expert system consists of three major parts: declarative knowledge about the problem, procedural knowledge about the problem solving method, and a control system or the inference engine.

The most important part of an expert system is the problem solving knowledge, also called procedural knowledge or domain knowledge. It is what makes the expert system able to solve the problem. This knowledge is specific to a problem domain, in our case, process planning. The problem solving knowledge contains both facts and rules. A typical fact in a process plan may be: "The diameter of drill DX0021 is .25 inch," or "Machine ML001 can perform drilling, boring, and reaming operations." Rules are quite often rules of thumb which human experts acquired over the years. These rules of thumb are also called heuristics. For example, most of the information presented in Chapter 4 is knowledge of this type. How good the knowledge is is crucial to the success of a process planning system. As discussed in Chapter 4, there are several ways to acquire the knowledge. However, in order to put it into an expert system, it must be represented in a usable form. How well the knowledge can be used depends on how well the knowledge is represented. The representation of the process planning knowledge is thus something that needs to be carefully studied.

Declarative knowledge contains mostly facts. These facts are facts about the problem being studied. In process planning the problem is how to make a part; thus, the facts about the problem are design details. The major issue here is the representation. This representation needs to be complete, unambiguous, efficient, and compatible with the problem solving knowledge.

Finally, the third part of the expert system—inference engine. The inference engine or the control system, contains the general problem solving knowledge. Part of an inference engine is an interpreter. It interprets the procedural knowledge and resolves the procedure embedded. The inference engine also has built-in search algorithms which determine the sequence in which rules are applied and facts found. The search method is called chaining. Since the inference engine is supposed to be general, it is not necessary to develop a special inference engine for a specific application. Many expert system shells provide either a built-in inference engine or tools to customize one's own inference engine.

Now that we have discussed the basic elements of an expert system, it is worthwhile to discuss briefly how an expert system works. The inference engine begins with a goal, say "make hole_1." *Make* is the operator and *hole_1* is an attribute. It starts searching rules applicable to make something. Based on the search algorithm, it found the first rule that is applicable to the task. The way it detects its suitability is by checking the conditions (antecedent) of the rule. A rule is in the form of *IF . (condition) . . . THEN . (action)*. Say, the first clause in the condition of rule_1 is "surface_type = flat surfaces." The inference engine retrieves the surface_type from entity "hole_1" and found it to be hole, the rule is not "selected" (fired). Then, the system goes on to find another rule, say, rule_2. It happens that rule_2 has a "surface_type = hole" clause. After satisfying all condition clauses in rule_2, the action clauses are executed. One action clause is "select boring_process" and the other is "diameter of hole_1 = diameter of hole_1 − 0.01." The boring process is thus selected. This rule also reduces the diameter of the hole to provide clearance for the boring process. These action clauses change the facts describing the part. Since the hole still exists, a new goal of making the smaller diameter hole is generated. The procedure repeats itself until the hole disappears. The sequence in which processes are selected in this example is a reversed sequence for actual machining.

In the above example we can see that the inference engine interprets the meaning of the clauses and takes action according to these clauses. It also applies its built-in algorithms to search among a large number of rules which should be applied next. The process planning knowledge is embedded in those rules. The rules tell under what kind of condition a process can be used and what changes it will make to the workpiece. Finally, the description of the part design as well the intermediate workpiece shape and technological characteristics are represented in a format compatible with the rest of the system.

In the following sections we will discuss in detail how to build a process planning system using expert system technology. Since there are numerous ways one can design a system, we will try to present different approaches for building each part of the system. An example expert process planning system will be presented in the next chapter.

5.1 Personnel in Building the System

Although the majority of expert process planning system built and referenced in this book were built by one or a small group of engineers, building a usable expert process planning system requires many more people. It is worthwhile to discuss just briefly the personnel needed for such an undertaking. We may classify the people around an expert process planning system into three categories: user, developer, and information provider. In expert system terms, they are user, knowledge engineer, and domain expert. The key player is of course the knowledge engineer. The knowledge engineer is the system analyst and the programmer. His or her job is to acquire knowledge from the domain expert and build the software using the knowledge. A knowledge engineer must be an expert in the programming tool (e.g., LISP, OPS5, KEE, etc.) used to build the expert system. The person(s) also need to know well the principles and techniques of building expert systems, such as knowledge representation methods. In the past it was believed by many software persons that a good knowledge engineer without any training in the problem domain could acquire enough domain knowledge through interviewing domain experts to build an expert system. However, the analogy is like asking a newspaper reporter to write a technical book. Obviously, she or he will be able to do an excellent job as long as the book stays at the training manual level.

Since the knowledge engineer selects the knowledge representation method and the problem solving approach, not having prior knowledge about the problem domain may lead the development in a wrong direction. In order to develop a successful expert process planning system, the knowledge engineer must have enough knowledge in manufacturing to begin with. More expert knowledge can be obtained later from domain experts. It is preferably to have a team of knowledge engineers consisting of manufacturing engineers and computer scientists. Every team member should have enough knowledge about building expert systems as well as knowledge about process planning.

The domain engineers are experienced process planners, manufacturing engineers, methods engineers or technicians, and machinists. They can provide first hand knowledge about the practice of the shop and information from other sources, such as handbooks and manuals. Since there are always variations in practice between shops, it is vital that local experts be consulted. Domain engineers should be well informed of the principles of expert system, the intention of the expert system project, the benefits and the limitations of it, and the consequences of having it. The cooperation of domain experts is essential to the success of the implementation. Since the very same person's job might be jeopardized by the system, to gain their trust and ensure them their future are most important.

Quite often the end users of the system are the same group of domain engineers. An expert process planning system does not always replace human planners. Often it is used to assist human planners. It helps humans do things better, faster, or both. It contains either the experience of the best human planner or the collective experience of a group of planners. Less experienced human planners can learn from the system.

Since current expert process planning systems cannot learn by themselves, the dynamic changes of the manufacturing system need to be captured and reflected in the knowledge base. A constant revision of the knowledge base is necessary. The source of the new knowledge is the system users. A well-known expert system XCON, used in Digital Equipment Corporation for VAX computer configuration, has kept a constant number of knowledge engineers to handle the update and addition to its knowledge base.

5.2 Expert Process Planning System Structure

As discussed previously, an expert system consists of three major parts. Figure 5.1 shows a block diagram of such an expert system. It also shows an additional user interface. Before we can start designing an expert process planning system, we first need to decide the mode of system operation. Quite often we take an expert system as an interactive system, that is a user consults the system interactively for advice similar to what one may get from a human expert. Actually, this need not be the case. An expert can also work on batch mode where jobs are submitted to him or her in a batch. Although a few prototype process planning systems do use an interactive mode, the majority of systems run batch mode. Unlike medical diagnostics or manufacturing control, in process planning the complete design information is readily available; therefore, little interaction is neces-

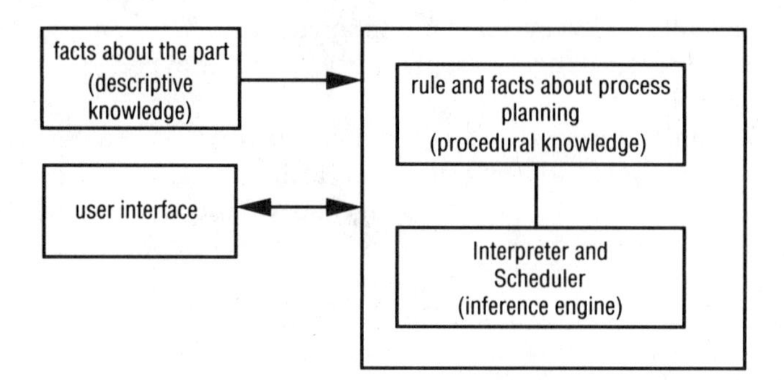

Figure 5.1 A basic expert process planning system structure

sary. It is easier to provide the system complete knowledge about the problem in a batch, than it is to interact with the system to answer questions. In the latter case, the process planning system works as a consultant to the human planner. It takes a long time to complete a planning session.

If the batch mode is selected, the user interface is reduced to provide a means to start the planning, display the results, and/or provide an explanation. Complete design information needs to be converted from the original data format into the declarative knowledge format. Since most design data bases provide only data, information needs to be extracted from the data. What information needs be extracted and prepared for the system has to be decided during the system design stage. Whether a user is responsible for translating the design data or whether an automatic interface will do the work also needs to be considered. When users are asked to do the translation, it may take a lot of time and effort to do so, thus reducing the effectiveness of the process planning system. However, a general method for automatic interface, as discussed in Chapter 3, is still not available at this time.

An ideal expert process planning system needs additional system functions, however. The explanation facility traces the reason a system concludes an answer. The rationale behind applying (firing) a rule is displayed. It either traces certain functions or reconstructs the reasoning steps. As in a traditional computer program, only simple tracing during the execution is available. The purpose of an explanation facility is twofold: debugging the knowledge and training the non-expert. Since either erroneous rules may be acquired from a domain expert (a process planner or a machinist) or a rule may not be properly entered into the system, the expert system may get incorrect answers. Without an explanation facility to help debug the knowledge base, it would be extremely difficult to pinpoint the error.

The purpose of having a knowledge acquisition module is to

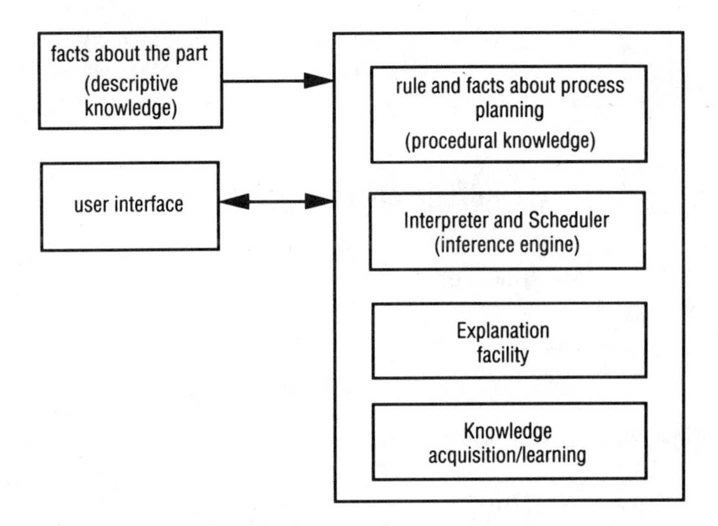

Figure 5.2 Ideal expert system structure

assist the user of the system to update the knowledge base, because we know that a manufacturing system is never static. There are constant changes in machines, tools, and fixtures. New manufacturing processes may be added to the system. Sometimes, through our better understanding of the process, we may want to modify some of the existing rules. Physically acquiring knowledge from human experts is exterior to the system and thus a different matter. The function of the knowledge acquisition module in the process planning system acts as the agent to assist in transforming external knowledge into internal knowledge representation. It needs to ensure that the new rules are consistent with the existing rules. If there is any contradiction of the newly entered rule with the existing rules, it needs to be detected by the module.

There are several aids used in knowledge acquisition modules, knowledge-based editor, explanation facilities, and knowledge-based revision [Buchanan, et al. 1983]. A knowledge-based editor helps to check the syntax and/or semantics of a new rule. It avoids grammatical and typological errors and make sure that the rule is consistent with the other rules in the knowledge base. The function of an explanation facility has been explained before. There are a few ways knowledge-based revision can be performed. The basic idea is to automatically run problems with the newly revised rules. If any conflict is detected due to the new rules, the user is informed.

Learning means a system acquires knowledge automatically. It is still a very difficult problem to solve. A small number of process planning systems (e.g., Tsatsoulis and Kashyap [1988]) are said to have a learning capability. However, the learning is very limited. For example, the system bases its learning on forming patterns. It is a method for learning about how to form part families instead of learning rules of process planning.

5.3 Planning Strategy and Inference Engine

The process planning task is a task of selecting methods (processes), resources (machines, tools, etc.), and determining the sequence of applying methods. An ideal expert process planning system has several strategies which can be employed to solve the problems. We will try to discuss each of the strategies.

5.3.1 Strategy Based on the Planning Level

a. Local Information Only

Under this strategy, planning is done feature by feature or surface by surface (feature will be used in this section to denote both volumetric features and surfaces). Only local information is considered during planning. As long as the global relationships of features are not used in the planner, it is considered to be in this category.

The planning process repeats itself for each individual feature (Fig. 5.4). The result of each planning session is an operation plan. A sequence of processes may be included in each operation plan for the feature being planned. The final process plan is the concatenation of all operation plans. The majority of the existing systems are of this type. Some systems do not determine the final sequence of operation plans; they can only be considered as operation planners. Some other systems use a simple heuristic to determine the sequence in which

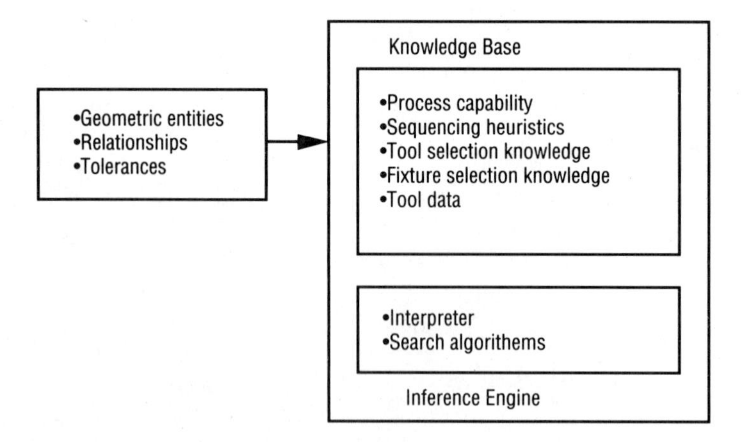

Figure 5.3 An expert process planning system

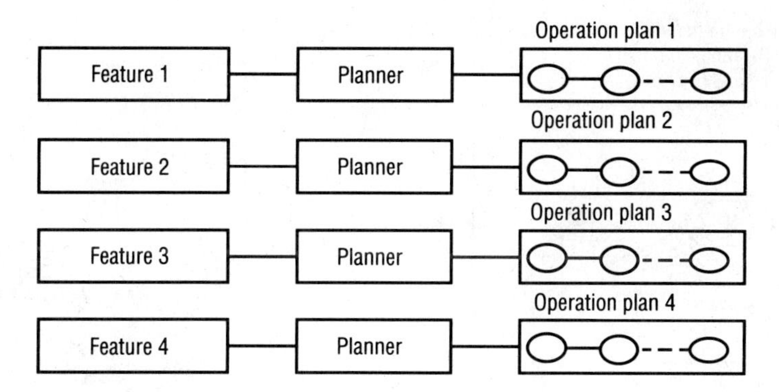

Figure 5.4 Local planning

features are to be machined. The operation plans are concatenated according to this sequence.

When this strategy is applied, the search tree is limited to processes for one single feature. As rarely more than three different processes are needed for a single feature and the available processes in a shop are few, the search tree is generally small and only a few levels deep. Even without any pruning the search should be completed quickly. Systems using this strategy are easy to build; however, the final process plan generated is not always correct or efficient.

TIPPS [Chang and Wysk 1985], TOM [Matsushima, et al. 1982], SIPP [Nau 1985], SIPS [Nau and Gray 1986], FirstCut [Cutkosky, et al. 1988] are but a few of the systems using this strategy. TIPPS can plan for a given setup. A graph of feature relationships is built based on feature contact relations and feature reference height. If two features contact each other, the one whose feature reference is higher is given the precedence. A simple heuristic uses the graph to determine the next feature to be planned. The planner is then used to plan the sequence of processes for the selected feature. During the rest of the planning session, no feature relations are used.

SIPP, SIPS, TOM, and FirstCut work on an individual feature following the sequence given by the designer. The final process sequence is therefore determined by the design sequence.

b. Global Consideration

The second strategy is to consider the global relationship of features during the planning (Fig. 5.5). In order to make reasonable plans, it is necessary to consider feature relationships during the planning time. Many examples can be found in Section 4.5.1 and Fig. 4.9. This strategy is much more difficult to implement than the one discussed before. More information needs to be modeled in part representation and knowledge on utilizing this information for decision making needs to be captured in the knowledge base. Since most of the global

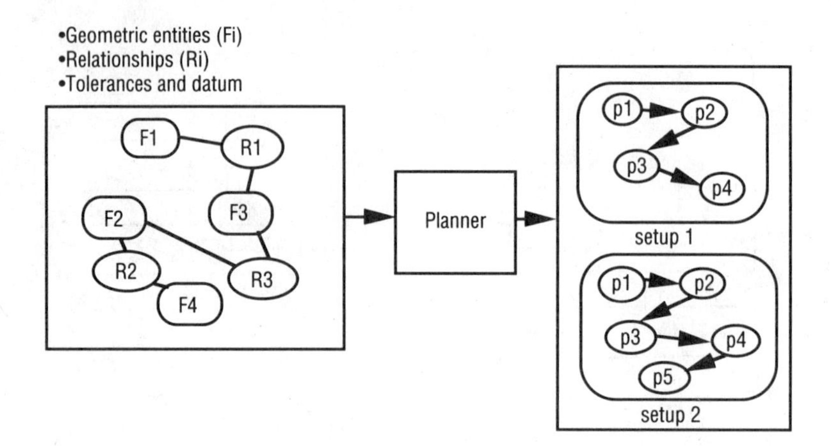

Figure 5.5 Planning with global information

considerations are based on geometric relationships, some kind of geometric reasoning mechanism needs to be built in. Just to mention a simple case, if the antecedent of one of the rules checks whether the current hole is too close to another hole, the question "too close" needs to be resolved. Procedure needs to be written to find all distances between the current hole and holes whose axis is parallel to the current hole. If one is found less than a given value, then the clause is said to be true. Since there may be many conditions to be checked, if all possible cases are checked, the planning can take an extremely long time to finish. An early decision needs to be made on the domain of the problem, thus cutting down the conditions.

Few of the existing expert process planning systems belong to this category. One example is a system called QTC [Chang, et al. 1988]. The QTC system deals with only a limited number of features. More details of the system design are presented in Chapter 6.

5.3.2 Based on Input Data View

a. Volumetric Feature-based

The basic geometric entity used in the system is a volumetric feature. A volumetric feature can be a hole, a slot, a pocket, etc. The geometric capability of a manufacturing process is described using the names of volumetric feature (Fig. 5.6). For example, a drilling process is described as a process which can produce holes. Drilling is selected when a hole is present in the design. Tools can also be selected based on the feature and its parameters and attributes. For example, drill diameter and length is selected based on the hole diameter and depth. The drill bit geometry is selected based on the hole bottom geometry. The hole bottom geometry is normally represented by some attributes attached to the hole and not represented explicitly.

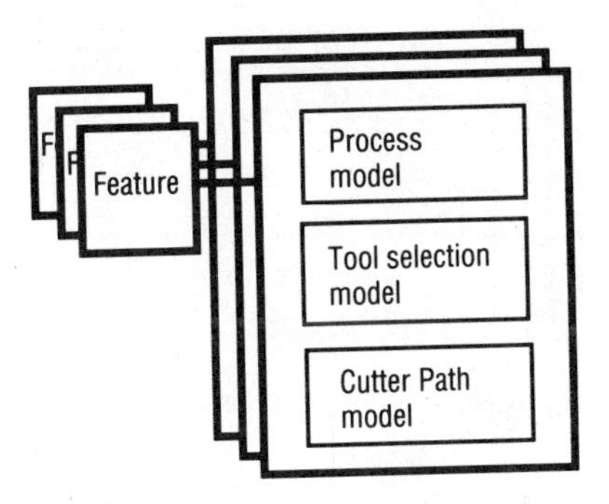

Figure 5.6 Features and associated planning models

Cutter path can also be generated based on a model for the feature. The final cutter path for the entire part can be either a concatenation of individual cutter paths or a merged and modified cutter path depending on the level of planning. Most of the existing expert process planning systems use volumetric features.

b. Surface Characteristic-based

The basic geometric entity used in the system is a surface. The focus in this case is the surface characteristics. Some surfaces are the by-product of a process applied to a main surface. Using the APT terminology, the main surface is the part surface, both the check and the drive surfaces are by-products of the operation. The main surface is normally used to represent the whole set of surfaces for process selection. The planner should not be confused by the extra surfaces. How to eliminate secondary surfaces from being planned can be difficult task for the system. Often, this is done by a human interface who inputs to the system only the main surfaces.

5.3.3 Based on Method

a. Composite Component

One way to build a process model is to use the composite components concept [Chang and Wysk 1985]. Although the composite components concept has been used in earlier non-expert process planning systems (e.g., CPPP [Dunn and Mann 1978]), it can also be used in expert process planning systems. A composite component is an imaginary component which contains all features (volumetric or surface) existing in a part family (see [Chang and Wysk 1985]. The composite component is parameterized and operation plans written for each feature using the parameters as variables. When used in an expert process planning system, rules are written for each feature and tool selection and cutter path models prepared. Figure 5.6 illustrates this concept. This approach is easy to implement yet it is less flexible.

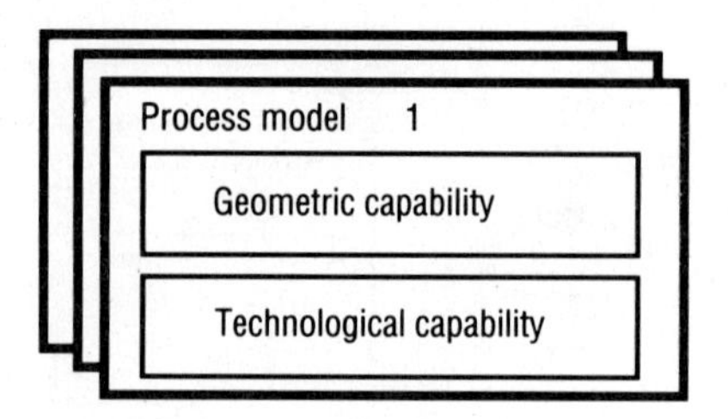

Figure 5.7 Generic process model

b. Generic Process Capability

The other strategy is to model the process capability based on an individual process. Instead of modelling what processes can be used for a feature, the rule determines which features can be produced by a process. When a new feature is developed, its name is inserted into the capability clause of processes that can produce it. It is most appropriate for solving process selection problems.

5.3.4 Based on Planning Direction

a. Forward Planning

Forward planning means the search of the planning process starts from a raw workpiece and tries to reach the finished part description (Fig. 5.8). It is the same as the actual machining process. The condition of a process rule is the raw workpiece shape and surface condition and the action is the operator which removes a feature frcm the workpiece. A simplified rule for a boring process can be:

IF (hole exists) THEN (hole diameter = hole diameter + 0.01″) and (surface finish = 20 μ inch).

Since the majority of machining processes can begin on a raw workpiece (boring is one of a few which requires an initial geometry to be present), a large search tree can be expected. As illustrated in

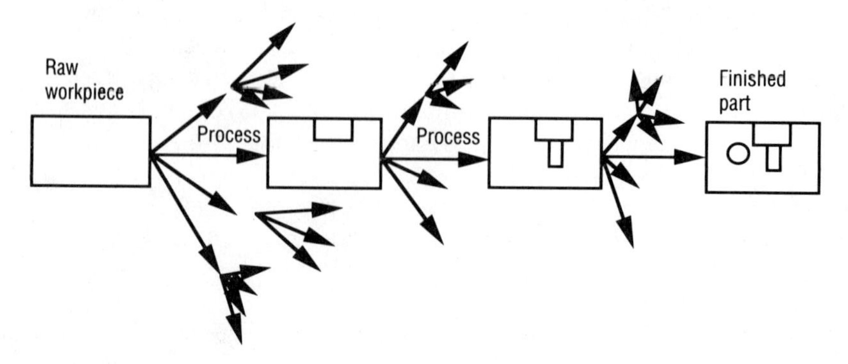

Figure 5.8 Forward planning

Fig. 5.8, in order to find the final goal, the combination of processes can be numerous. Pruning is essential to make the problem tractable.

b. Backward Planning

Backward planning takes a reverse process. It starts the planning from the finished part and tries to fill back to the raw stock state (Fig. 5.9). One strategy is to use the generate and test method combined with the backward planning strategy. At each iteration the current state of the workpiece is compared with the finished part (goal). A set of features results from this comparison. Among the features in the difference set, one is chosen based on the sequencing rule. Processes are searched, and the one which is most appropriate for the operation is selected. The antecedent of the rule is the finished state a process can achieve and the consequent is the initial state. The features are filled back to the raw workpiece state (Fig. 5.10).

It is obvious that the backward planning strategy makes the planning much easier. Since the antecendent of a process rule describes what may result from applying the process, it can match with the design well. Figure 5.10 illustrates a process of applying backward planning on one hole. The surface finish specification of the hole is 20 μ inch. After failing several drilling processes, a boring process is selected. The rule for the boring process is:

IF (the feature is a hole), (surface finish ≥ 20 μ inch) THEN (select boring process), (tool diameter = hole diameter), (hole diameter = hole diameter − 0.01″) and (surface finish = 500 μ inch).

Instead of enlarging the hole, the consequent of the boring process actually reduced the hole size by an amount needed as the clearance. The surface finish is also relaxed. The rule is interpreted as follows: If the required surface finish is worse than 20 μ inch, the hole can be bored. However, before the hole is bored, it must be 0.01″ smaller than the desired final hole size. The surface finish before a boring process can be used need not to be too good (represented by surface finish − 500 μ inch).

After the current state of the hole has been modified, a drilling process is selected. Finally, the hole disappears. The drill diameter is

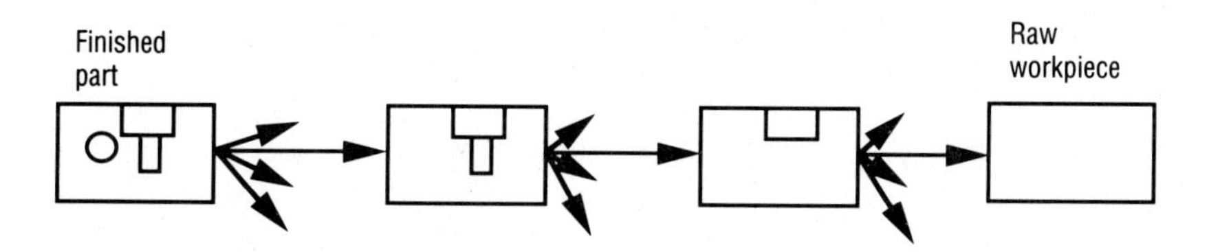

Figure 5.9 Backward planning

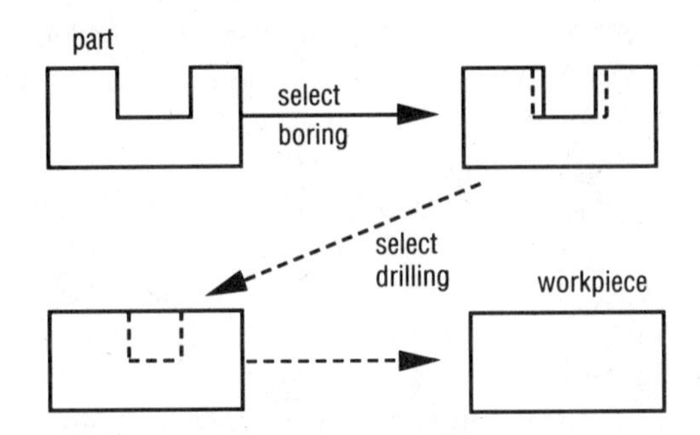

Figure 5.10 Backward planning on a precision hole

the current hole diameter which is 0.01" smaller than the designed hole diameter. This works well because the appropriate tool diameter can also be found. As a matter of fact, each of the intermediate workpiece states is found. The final process sequence is the reversed search sequence.

5.3.5 Direction of Chaining

Chaining is a term used in an expert system but not in process planning. There are backward chaining and forward chaining. Chaining it is quite often confused with the direction of planning. Forward chaining means rules arc matched with facts to establish new facts [Waterman 1986]. In the opposite, backward chaining starts with a goal and tries to establish the facts it needs to prove that goal. Figure 5.11 illustrates the chaining operations. In forward chaining, the facts are matched with the antecedents of rules. In the example, the rules are searched in a sequential manner. Since the antecedents of Rule1 does not match, it is skipped. The antecendents of Rule2 match with the facts; it is said to be fired. When a rule is

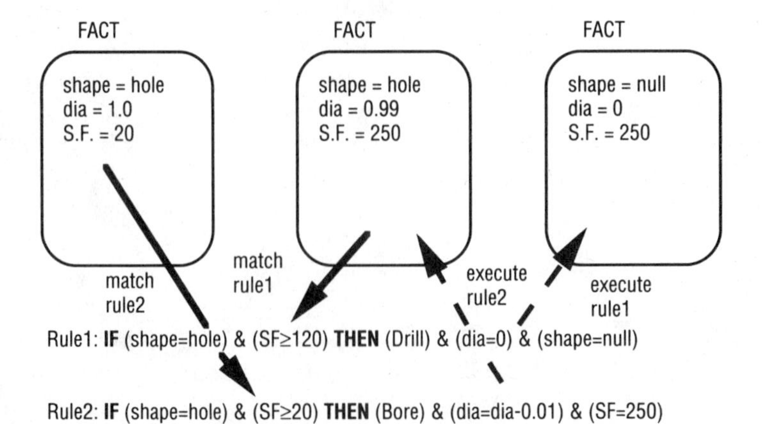

Figure 5.11 Forward chaining

fired, the consequents are executed. The execution results are place in the fact list (blackboard). This process is repeated untill the final goal is reached. The final goal in this case is to make the shape null.

Backward chaining, on the other hand, detects that one of the antecedents of Rule1 is not in the fact list. The system then tries to establish it as a fact. In order to accomplish this, the rule which will make it true is searched. In the example, Rule2 has a consequent which satisfies this requirement. To execute this consequent, the antecedents of Rule2 must be true (in the fact list). Fortunately, we found both clauses in the fact list; therefore, Rule2 can be fired.

Step 1: Begin with the antecedent of Rule1.
 (SF ≥ 120) is not in the fact, try to establish it.

Step 2: The clause (SF≥120) found in the consequent of Rule2.
 Both (shape=hole) and (SF≥20) satisfy.

Step3: Fire Rule2 to obtain (SF≥20)

Step4: Fire Rule1 to obtain the goal.

Both chaining methods may be used in an expert process planning system. In fact their use may be mixed in a system.

5.4 Declarative Knowledge About the Part

The description of the part is the input to an expert process planning system. Generally, the representation must be complete and unambiguous. In order to make the system efficient, the representation should also carry only the minimum amount of information necessary. Information not needed for process planning should not be passed to the process planning system.

There are internal and external representations. Distinguishing between internal and external is based on the process planner's prespective. The original part design, either on a CAD system or on a engineering drawing, is the external representation. After a conversion process, the external representation is translated into a data format and fed into the process planner. This converted data representation is usable directly by the process planner and is called the internal representation. The external representation has been discussed in Chapter 2. The conversion process can be either manual or automated. The automated conversion process converts a CAD data file into process planner input. It is called CAD interface or geometric reasoning. In Chapter 3 several of the existing CAD interface methods were discussed. In this section we will discuss the internal representation of a part design. There are two ways such data can be input into the process planner, either through an interactive session or through a batch process. For any realistic part, the amount of information necessary to describe the part is prohibitive for the slow interactive process. Therefore, the batch process is recommended.

The part information necessary for process planning includes geometry, geometric relationships, dimension and tolerances, and additional manufacturing specifications. Depending on the sophistication of the planner, all or part of the information is included in the representation.

For part internal representation, Nau [1983] identifies Frames, First Order Predicate Calculus (FOPC), and Semantic Nets as being suitable for representing the part. Nau [1986] suggested using Frames for part representation. The QTC system [Chang, et al. 1988] also use frames for part representation. Hummel and Brooks [1986] used 'Object' representation in their XCUT system. Inui, et al. [1987] reported a system, XMAPP, which uses a combination of first order predicate logic and an object oriented approach. The GARI system mentioned before uses first order predicate logic as well. There are many other informal methods that have been used in different process planning systems.

5.4.1 Frame

A frame system is a network of nodes and relations organized in a hierarchy. A frame is a data structure which represents an entity type. A frame consists of several named slots. Each slot can store values, pointers to other frames, or a procedure (Fig. 5.12). When a slot is filled, it is said to be instantiated. The slot value can be restricted to a certain range or from a set of values. For example, the diameter of a hole is restricted to a range which is machinable by the shop. Values outside the range will be rejected. A slot may have a default value. When the slot is not filled, the default value is used. Default values can be used for tolerances. Since tolerances are specified for only surfaces that need special attention, tolerances for other surfaces can be defaulted to a preset large value. This feature of the frame system relieves the user from entering every slot value. One of the important features of the frame is the inheritance. A structure of frames may be set up. In the structure both parent/child link and sibling link are allowed. Part of the slots can be inherited by the child from the parent. For example, both thru slot and blind slot belong to class slot. The property of slot, such as dimension, dimensional tolerance, position, positional tolerance, surface finish, etc., can be inherited by both sub-classes. One frame can also be linked to another frame by a pointer stored in a slot.

Procedures or functions can be attached to a slot. The procedure may be executed when a) new information is placed in the slot, b) information is deleted from the slot, c) information is needed in the slot. This procedure is also called demon.

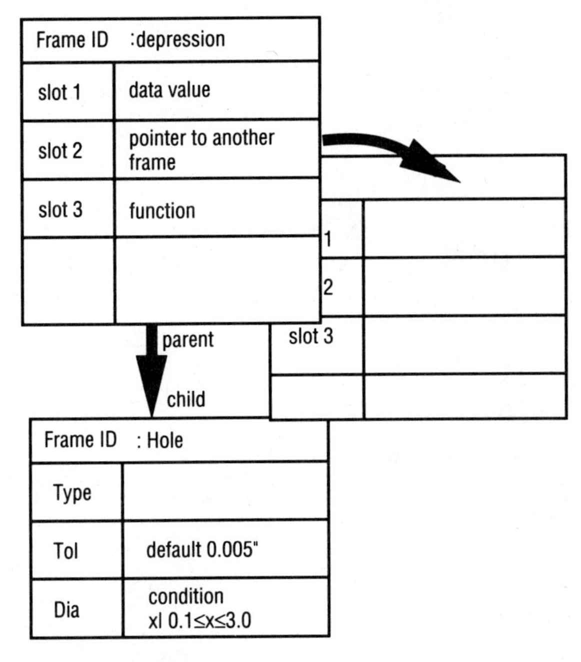

Figure 5.12 A frame

5.4.2 Manufacturing Feature Information

Although parts can be represented in many different ways, for process planning purposes the easiest way to model the machined surfaces is through manufacturing features. The manufacturing features are different from design oriented features. The former are related to the trace of cutting tools and the latter may be related to function or special geometric shape. Manufacturing features can be classified as primary features and secondary features. Following is a partial list of manufacturing features.

- flat surface
- blind slot
- thru slot
- pocket
- V-slot
- T-slot
- Y-slot
- dovetail slot
- step

- blind step
- plain round hole
- threaded hole
- taper bore hole
- curved slot
- non round hole
- island
- sculptured surface cavity
- convex sculptured surface

Some secondary manufacturing features are the following:

- counter bore
- counter sink
- flat bottom
- conical bottom
- edge chamfer
- chamfer
- radial groove
- face groove
- keyway
- splines

Each feature has its own parameters to define the dimension of the feature. For example, a round hole can be defined by its diameter and length, pocket (flat bottom polyhedral pocket) can be defined by the numbers of sides, vertex coordinates, and depth. In order to define the position and orientation of a feature, additional information such as reference location and orientation vector need to be included. For process planning both the dimensional and geometric tolerances need to be included too. Dimensional tolerance can be attached to the dimension. Some geometric tolerances, i.e. roundness, flatness, cylindricity, and straightness, refer to the feature itself and can be attached to the feature as attributes. Positional tolerance can be defined by a vector from the position reference to the feature reference. Other geometric tolerances, i.e. perpendicularity, parallelism, angularity, line profile, surface profile, concentricity, circular runout, and total runout, require one or several datum surfaces. In order to define them, an explicit surface (feature is volumetric entity) definition is required. This means that a mixture of volumetric features and surface features have to co-exist in the representation. The representation becomes extremely complicated. There is yet an acceptable solution available.

5.4.3 Representation Using Frames

In this section, a few actual implementations of frame representation are illustrated. These systems are by no means the ideal systems. Nevertheless, they serve as good examples.

a. SIPP aand SIPS

Nau used frames as the representation scheme in SIPP [Nau 1985] and later SIPS [Nau 1986] systems. He proposed a feature hierarchy in which features are classified into surface and contained-surface. Surfaces (actually volumetric features) are classified into cylindrical and non-cylindrical surfaces, etc. In the lowest class in the hierarchy are features such as hole, straight-slot, flat-surface, etc. This hierarchy can be mapped naturally into a frame hierarchy with property inheritance. The slots depend on the type of the feature. SIPP and SIPS generate plans for individual features; only one feature is being considered at a time. This representation is sufficient for the purpose. The following example shows the frame system used in SIPP.

```
type(surface, feature).
slots(surface,[
    [surface_finish,X,number(X)].
    [contains,X,list_of_atoms(X)].
    [adjacent,X,list_of_atoms(X)]
type(hole,surface).
slots(hole,[
  [norm,[X,Y,Z],(number(X),number(Y),number(Z))],
  [depth,X,number(X)],
  [diameter,X,number(X)],
  special_features,X,X is_in [none,straight_chamfer,inner_fillet,out_fillet]]
])
```

Since SIPP is implemented in PROLOG, the syntax follows the PROLOG syntax. Upper case letters represent variables. SIPS uses LISP as the implementation language. The general stucture is similar.

The type statement defines the hierarchy, i.e. hole is a sub-class of surface. Slots of surface are inherited by all children frames, such as the hole frame. For example, surface finish is a property of any surface; therefore, it is defined in the surface frame. The data value restriction is set by the 'number' and 'is_in' functions. An instance of a hole frame can be the following:

```
item(h,hole).
slotvals(h,[
  norm=[1,0,0],
  surface_finish=64,
  depth=1.5,
  diameter=0.5,
  special_features=none
])
```

There is no slot for tolerance and feature relationship other than contain-in relation. The lack of global information restricts it as a full process planning system.

b. GEFPOS System

The GEFPOS system [Vissa 1987] classifies a part in a hierarchical structure. The major classes are features and tolerances. In order to attach tolerances to the volumetric features, face features are also included in the representation. Features are classified as machined and un-machined features. The un-machined features may be used for fixturing purpose. The classification is richer than that of SIPP and SIPS. A frame system is used to represent the part structure. Most of the information necessary for manufacturing is included. For example, the top level frame part is represented in the following frame.

```
(T001
  (A_KIND_OF (VALUE PART))
  (PART_NAME (VALUE EXAMPLE_1))
  (BATCH_QTY (VALUE 1))
  (ORDER_QTY (VALUE 1))
  (MATL_SIZE (VALUE (3.2 2.2 1.2)))
  (MATL_TYPE (VALUE STEEL))
  (MATL_SPECN (VALUE SAE_1050))
  (MATL_COND (VALUE HARDENED_AND_TEMPERED))
  (GENL_HARDNESS (VALUE (450 500)))
  (SURFACE_TREAT (VALUE NIL))
  (MADE_UP_OF (VALUE (F2 F5 F7 F8 F10 F11 F13 F20 )))
)
```

As can be seen, production quantity and material specifications are included in the part description. The last slot stores all the top level features contained in the part. For each individual feature, data, such as dimensions, tolerances, approach direction, location, etc., are stored in the slots. For example, the following is a instance of a hole.

```
(H1
  (A_KIND_OF (VALUE PLAINHOLE))
  (DIAMETER (VALUE .6))
```

```
    (DEPRH (VALUE 1.0))
    (AXIS_DIRECTION (VALUE (0 0 1)))
    (APPROACH_DIRECTION (VALUE (0 0 -1)))
    (DEPTH_REF (VALUE F10))
    (ENDS_IN (VALUE F18))
    (BOTTOM_TYPE (VALUE NIL))
    (X_REF (VALUE F17))
    (Y_REF (VALUE F16))
    (X_DISTANCE (VALUE 1.0))
    (Y_DISTANCE (VALUE 1.0))
    (SECONDARY_FEATURES (VALUE (CB1)))
    (CONTAINS_FEATURES (VALUE NIL))
    (CONTAINED_IN (VALUE F10))
    (DIAMETER_TOL (VALUE .001))
    (DEPTH_TOL (VALUE NIL))
    (SURFACE_FINISH (VALUE 32))
    (POSITION_X (VALUE .005))
    (POSITION_Y (VALUE .005))
    (CIRCULARITY (VALUE NIL))
    (CONCENTRICITY (VALUE NIL))
    (CONCENTRICITY_REF (VALUE NIL))
    (PERPENDICULARITY (VALUE NIL))
    (PERPENDICULARITY_REF (VALUE NIL))
    (HARDNESS (VALUE 500))
)
```

In the same frame three categories of information are stored. They are information about the feature itself, information about the relationship of the feature to other features, and information used for matching the feature to the process. The reference slots, such as, 'x_ref', 'perpendicularity_ref', etc., define the data for a tolerance specification. They also include a slot on surface hardness. This information is important, since in many parts local heat treatment is required. Process sequence and tool selection are affected by the surface hardness. This representation is implemented in COMMON LISP.

c. QTC System

The QTC system [Kanumury 1988, Chang, et al. 1988, Kanumury, et al. 1988] uses the same approach in part representation. In addition to the characteristics mentioned in the previous two systems, the QTC representation contains more feature interaction and approach information. The frame for a slot is shown below.

```
(slot_1
  ((id 2)
  (type 1)
```

```
    (feat_type 0)
    (length 4.50000)
    (width 0.80000)
    (depth 0.500000)
    (b_type FLAT)
    (handle_face_ref 0)
    (ref_feat work_piece)
    (ref_face face1)
    (surface_finish 250.000000)
    (length_tol_range (-0.050000 0.050000))
    (width_tol_range (-0.050000 0.050000))
    (depth_tol_range (-0.050000 0.050000))
    (locational_tol_range (-0.086603 0.086603))
    (parallelity_tol_ref ())
    (perpendicularity_tol_ref ())
    (angularity_tol_ref ())
    (approach_direction ((0.000 0.000 -1.000)))
    (aux_approach_direction ())
    (feed_direction ((0.000 1.000 0.000)
      (0.000 -1.000 0.000)))
    (thru_direction ((0.000 1.000 0.000)))
    (thru_type THRU_ONE_END)
    (contains_b (hole1))
    (contains_s ())
    (contained_b (slot1))
    (contained_s ())
    (merge ())
    (split ())
  (causing_split ()))
```

The additional information is inferred by a geometric reasoning system. Details of the approach will be discussed in Chapter 6. The unique part of the system is that it actually keeps three representations during the planning time. The frame representation is used for most of the planning tasks.The original feature-based CSG model and a evaluated B-Rep model are also used for fixturing planning and cutter path generation. Both the fixturing planning and cutter path generation are mixed in the process planning system.

5.4.4 Other Representations

A few other representations have also been used in expert process planning systems. Inui, et al., [1987] showed a system using first order predicate logic to represent a part. In their system a form feature is an entity and is represented by predicate; relationships between the features and its component face elements are defined by binary predicates. For example, a hole 'h1' with entry face 'f5' is represented:

```
hole(h1)
entry_face(h1, f5)
```

Relations between features are also defined by some predicates, such as branch_connect. Most of the relations are derived using rules. Dimension of a feature is defined by a 'distance' predicate, for example,

```
distance(f1, f5, 0.5)
```

Normal, diameter, tolerance, etc., are treated as attributes to an entity. Attributes are handled by a object oriented language, FLAVOR. The rest of the representation is written in a PROLOG-like language. Since the representation is rich in geometric definition, geometric reasoning can be done directly. There is a strong similarity between this representation and the one used in [Henderson 1984, Henderson and Anderson 1984].

One of the earliest expert process planning systems, GARI describes a part using three elements: feature description, dimension description, and general information. Features are described by their type, starting-from and opening-into faces, dimension, surface finish, etc. The information explicitly represented in other systems is implied in the starting-from and opening-into clauses. An example of a hole description is shown below:

```
(H3 (type countersink-hole) (diameter 5)
  (countersink-diameter 7)
  (starting-from F11) (opening-into F12)
  (quality rough))
```

Dimension between features and dimensional tolerance are represented in a predicate, such as, (distance F1 F2 10 ± 0.001), (concentricity H1 H2 ± 0.01), etc. Material type and overall quality are also represented in predicates, such as, (matter SAE-1050), (quality 6.3).

A system Propel [Tsang 1987] that is evolved from GARI has a similar kind of representation. However, Propel uses a feature hierarchy and a relations hierarchy to define the part. In the feature hierarchy, a simple feature is classified into planar, cylindrical, cubic, profile, and others. Each class is then further classified into sub-class. For example, cylindrical has bore and shaft—two sub-classes. Relation hierarchy includes positioning, dimensional, and form—three classes. They are then further classified. An example of Propel part model follows:

```
(part support (Ra 6.3)(position-tolerance js)(s-x 148)
  (s-y 75)(s-z 105)
  (quality 10))
(feature bore1 (type bore)(normal z)(diameter 18 H8)
  (quality 8)(Ra 6.3)
  (rough-state solid)(opens-onto face2 face4)(s-y 18)
  (s-z 63))
(coaxiality borel B1 1.2)
(distance y bore1 face5 58 .05 -.05)
(parallelism bore1 face5 .06)
(distance z bore1 face1 21 .1 -.1)
(feature face4 (type circular-face)(normal z-)
  (Ra 6.3)(support-of bore1)(s-x 35)(s-y 35))
(distance z face4 face2 63 .3 0)
(parallelism face2 face4 .04)
(perpendicularity face1 face5 .05)
(distance y face5 F1 16 .8 0)
(distance y face5 F2 27 1.6 0)
(rough F5 (s-y 16)(s-z 92)(normal x-)(type planar))
```

In this representation, geometric tolerances and data surfaces are modelled using predicates. This allows much more detailed part information to be represented. However, the preparation of these data can be a very tedious job. In general, a representation should provide enough information for the planner to use. The format should be easy to retrieve and to manipulate.

5.5 Procedure Knowledge of Planning

In a planning system, depending on the sophistication of the system capability, the planning knowledge may do one or several of the following things: select processes, select cutting tools, select machining parameters, select machine tools, select fixturing methods, sequence the operation, and generate tool path. Not every task is suitable to be performed using the same approach. Some tasks are dependent on others. The selection of cutting tools can only be done after a process has been selected. The machine tool can not be selected unless the processes, tools, and parameters have been decided. The tools and parameters are used to calculate power and force requirements that are essential for machine selection. Tool and machine tool selection can be treated as sub-task under process selection. Machining parameters selection depends on the tool selected. Tool path generation requires information on tool and machining parameters. The operation sequence on the feature level is not affected by other tasks. However, the sequence on the part level is affected by the tools and machines.

Most of the systems developed so far deal with only one or two of the above tasks. Often the sequence of operation on the part level is predetermined by the designer. Some systems use simple heuristics to determine the feature sequence. In this case, the process sequence is the same as the feature sequence. The process knowledge need not be too complicated. Most of the knowledge deals with only one process. The interaction between processes, tools, machines, and features is not considered. In order to consider sequence on the part level, a higher level sequencing determination knowledge needs to be included. The goal of the sequence should be minimizing the tool changes and machine changes.

The most commonly used representation method for procedure knowledge is rule. Although both rules and frames can be used. The majority of systems developed so far use rules to represent this problem solving knowledge.

5.5.1 Rule

Rule is in a form of **IF** (antecedents) **THEN** (consequents). The state of antecedent clauses are logically ANDed, if the result is TRUE, the consequents are executed. For process selection using backward planning (all examples in this section show the backward planning approach), the antecedents contain feature type, dimension range, tolerances, surface finish, etc., information. This information is also called process capability as discussed in Chapter 4. The consequents describe action that needs to be taken after a process is selected. A simple rule written in a pseudo language for a drilling process follows:

```
if((( H?type is hole)
  (H?diameter ≤3.0)
  (H?diameter ≥.1)
  (H?l/d-ratio ≤8 )
  (H?pos-dia-tol ≥0.003 )
  (H?neg-dia-tol ≥0.001 )
  (H?true-position ≥0.0004 )
  (H?roundness ≥0.004 )
  (H?surf-fin ≥100 ))
  then(
  (select twist-drill)
  (H?diameter = 0.0 )
  (delete H? )))
```

H? represents the feature currently being planned. The process capability data can be found in Table 4.5. In Table 4.5, it is shown that tolerance capability is a function of hole diameter. In the above rule and in most of the existing process planning systems, this is simplified. To actually do it right, a relational data table needs to be included in the rule.

The action clauses in consequents modify the state of the part (or feature). For a finishing process, such as boring and reaming, the action should relax the tolerances in the current state and reduce the dimension of the feature to provide clearance for the finishing operation.

Examples of rule representation for TIPPS and QTC systems have been shown in Chapter 4, so, they will not be repeated here.

GARI and Propel also use rules for process planning knowledge. In the above discussed systems, the weight of a rule is implied by the sequence the rule is placed in in the knowledge base. GARI and Propel include a weight for conflict resolution. For example sequencing rules are included in the knowledge base. For example, a rule in GARI determines the sequence between two cuts for two adjacent features:

```
(is-a &x hole) (is-a &y hole) (open-into &x &y) (not open-into &y &x)
=> (9 (before (roughing-cut &y) (roughing-cut &x)))
```

The rule can be interpreted as, if hole &x open into hole &y then with 90% confidence rough cut &y before rough cut &x. When two conflicting pieces of advice are concluded, the one with higher weight is selected. The assignment of the weight is based on an educated guess.

There are two ways to determine the process sequence. One is that used in GARI, using sequencing rules. The other approach is to use a higher level control structure to sequence the features, as is used in the QTC system. In the latter case, the cut sequence is determined by merging the cut sequence for each feature and the feature sequence. The objective is to minimize the tool changes. It is easier to obtain an optimal sequence using the second approach. Most of the current systems plan a sequence only for an individual feature. Any optimization of tool selection or sequence determination on the feature level is localized and won't guarantee a final optimal process plan.

5.5.2 Frame

As discussed in Nau [1986], using a hierarchical frame for process knowledge representation reduces the search space. Since knowledge is classified, during the search only the part of the knowledge which is relevant to the feature needs to be searched. When processes considered in the planner are few, this structure may not gain much. However, when a large number of processes are considered, this approach will prove to be efficient. The advantage of a frame representation is that it is a well organized structure. This can also be a disadvantage when it is used in representing process knowledge. As long as frames are used to represent technology capability of processes, it is adequate. When geometric relations are considered in the planner, it will be difficult to represent them in such a way. Readers may try to represent the knowledge given in Chapter 4, Section 4.5.1. Two systems using frame for process representation are introduced in this section.

a. SIPP & SIPS

Nau [1986] proposed to use a hierarchical frame for both part, process, and tool representation. The part representation has been shown in the previous section. The hierarchy of process knowledge representation is similar to that of part representation. A classification of processes based on shape capability is used for the hierarchy. The top level classes are surface process and hole process. Under each class there are surface/hole create, improve, and feature processes. Surface/hole create processes are processes which can begin on a blank. Surface/hole improve processes are finishing processes. Surface/hole feeature processes are processes for secondary features such as chamfering, threading, etc. Some machining processes are also subclassified based on their tool classes. For example, milling is classified into face milling, peripheral milling, end milling, etc. In case a process is used in roughing and finishing two stages, it is further classified into roughing and finishing processes. The tool hierarchy is similar to the process hierarchy except it is further classified by tool materials.

A process frame includes slots on relevant (feature machinable), cost, precost, restrictions on diameter, depth, and tolerances, and actions. Cost for a lowest level frame, such as a twist drilling frame, is an arbitrarily assigned weight. It represents the desirability of choosing the process. When the process has sub-classes, then the cost is the minimum of children's cost. Pre-cost is a slot used to record the cost for the roughing process. For example, before the reaming process a drilling process is used, the pre-cost is therefore the cost of the drilling process. The costs are use in a branch-and-bound search algorithm for selecting minimum cost sequence of processes. However, the sequence is limited to one independent feature at a time. The action slot may create a subgoal. The goal is to make a feature, the subgoal will be making a modified feature. The modification reflects what the selected process can do. For example, a boring process will create a subgoal which is to make a smaller and less accurate hole. The control mechanism (inference engine) applies the exact same set of rules to find a solution. Finally, when a roughing process is found, the action flags a success.The search stops when it is successful.

Frames for a hole-process and a twist-drill process in SIPP are shown below.

```
type(hole_process, process).
relevant(hole_process, hole).
defaultvals(hole_process, [ cost=1, pre_cost=0]).
restrictions(hole_process, H) :-
  H?contained_in eq Surface,
  Surface?type eq flat_surface,
  H?norm eq Hvector,
  S?norm eq Svector,
  parallel(Hvector,Svector).
type(twist_drill, hole_process).
defaultvals(twist_drill, [precedence=10]).
restrictions(twist_drill, H) :-
  H?special_feature eq none,
  H?diameter gte 0.0625,
  H?diameter lte 2,
  H?depth lte 6,
  H?surface_finish gte 100.
actions(twist_drill, P, H) :- success(P).
```

b. GEFPOS System

Similar to SIPP and SIPS systems, GEFPOS uses a hierarchical process structure to classify machining processes. Surface processes are classified into basic, sizing, secondary, and finishing processes. The classification is based on the progressive refinement in the capabilities of these processes. In the case of hole generating processes, in addition to the above mentioned four classes, there is a hole preparation process class. Since slots of a process frame are used to match with the slots of a part description frame, the number of frames is reflected by that used in part representation. An example frame for a jig grinding process is shown below.

```
(
Jig_grinding
(a_kind_of
  (value hole_finishing_process surf_finishing_process))
  (frame_label (value jig_grinding))
  (tool_type (value (mounted_point)))
  (required_machine (value (jig_grinding_mc)))
  (machined_features
      (value (plainhole taperbore island nonround_hole
                pocket blind_slot blind_step)))
  (batch_qty (value 1))
```

```
    (order_qty (value 1))
    (hardness (value 800))
    (width (value (0.04 2.0)))
    (l_by_d_ratio (value 3))
    (surface_finish (value (17 32)))
    (dimension_tol (value 5.0e-4))
    (individual_tol (value 1.0e-4))
    (related_tol (value 5.0e-4))
    (pre_machined_feature (value same))
    (pre_surface_finish (value 33))
    (pre_dimension_tol (value 0.005))
    (pre-individual_tol (value 0.005))
    (pre_related_tol (value 0.005))
    (pre_hardness (value 200))
    (finish_allowance (value 8.0e-4))
)
```

A matching predicate is used to match the conditions specified in the frame with the that specified in the part description frame. The slots with prefix 'pre' are action slots. The control scheme of this system is less sophisticated.

5.6 Other Issues in a Process Planning System

5.6.1 Process Parameters Selection

Process parameters for machining include feed and cutting speed. The selection of process parameters depends on the tool material, workpiece material, tool diameter, and depth of cut. It is mostly done through table lookup. Since there are very large number of materials, diameter and depth of cut combination, the storage of process parameters requires a large database. The standard METCUT Machining Data Handbook consists of more than a thousand pages of tables. It is not efficient to store them in the knowledge base. A relational database will be most appropriate for this purpose. Rules can be written to synthesize queries which will be sent to the database. The query will return the suggested feed and speed from the database. Before it can be done, a tool must be selected. The initial depth of cut can be selected based on the tool and workpiece material. Rules also need to be written to select among pairs of feed and cutting speed.

5.6.2 Plan Optimization

It is always desirable to find an optimum solution for a given problem. In process planning the goal is to find a set of operations that is least expensive, yet still can satisfy the part specification. When one takes a narrow view of the optimization problem, say, on tool selection, one can always build a mathematical model to optimize the cutting time or cutting cost. However, the problem is not as simple as it seems to be. Many factors affect the economics of manufacturing. When selecting a tool based on its machining, cost may seem reasonable. However, this optimal tool may mean extra tool inventory. Tool costs often are higher than the machine cost. When each feature is machined by an optimal tool, there may be a large number of tools needed for machining one part. This large number of tools then creates a series of problems in tool inventory, scheduling, etc., problems. The original intention, reduce cost, has failed. Reasonable goals are, therefore, to reduce the number of cutting tools used, reduce the number of setups, reduce the number of machines used, and reduce the number of tool changes.

5.6.3 Tool Selection

There are several decisions to be made in tool selection. Variables on a tool include tool type (specific kind of tool), tool geometry (nose radius, rake angles, clearance angles, etc.), tool material (HSS, carbide, hardness/grade, etc.), tool dimension (diameter, length, etc.), and/or tool assembly (insert, holder, etc.). Tool type is determined by the process type and is probably the easiest to decide. The rest of decision has to be made based on the workpiece material, hardness, feature dimension, and feature approach clearance. Rules for tool selection should include a rule for clearance checking. In order not to optimize the process plan, it is desirable to minimize the number of tools used. For milling, a rule should be written to use the same tool for several features as long as the difference among optimal tool sizes for those features are small.

So far most of the systems which do select tools consider only tool type and tool size. The QTC system takes into consideration all the factors mentioned here. In QTC the available tool data are stored in a relational database, tool selection rules generate queries to retrieve the desirable tools. Due to the number of tools a shop may carry (in the thousands), it is more desirable to have them stored in a database instead of in the knowledge base. This database should be integrated with the tool management database.

5.6.4 Machine Selection

Machine selection can be treated in the same way as tool selection. Since machines have fixed characteristics, they can be represented in a frame structure. The number of machines in a shop is limited, so the machine knowledge base won't be too big to be included in the planner.

To characterize a machine we will need at least the following slots: machine type, process type, work table size, work envelop, spindle horsepower, spindle taper type, feed and speed range, axis torque, accuracy, cost, etc. These capabilities should be matched with the requirements. The required horsepower and force need to be calculated using formulae introduced in Chapter 4. In order not to overload the best machine, it is important that machine loading be considered during the machine selection phase.

5.7 Tools for Building Expert System

There are many tools which can be used to build expert system. In the expert systems used in industry today, a wide range of tools is used. Although the majority of them use so called AI languages (symbol manipulation languages), such as LISP and PROLOG, every imaginable computer language has been tried. Some of the languages are easier than others when used in building expert systems. As a by-product of building expert systems, some expert system shells have been developed. An expert system shell is an expert system with its knowledge stripped. Since the system structure is already built, the builder of an expert system need only insert the knowledge. However, expert system shells are not flexible due to their fixed structure and inferencing strategy. Expert system building tools were developed to ease these problems. An expert system building tool is a set of utility functions which enables users to customize an expert system according to the specific need. A better user interface environment is also provided to the user. Following we will discuss the characteristics of a few popular languages and shells.

5.7.1 Programming Languages

As mentioned above, many programming languages have been used in building expert systems, e.g., TOM [Matsushima, et al. 1982], an expert process planning system, uses Fortran; TIPPS [Chang 1982] and XPLANE [van 't Erve 1988] use Fortran to build a rule-based language for process planning. However, the builders of a majority of the expert systems chose instead either LISP or PROLOG languages. Since building expert systems requires symbol manipulation, it is natural that a symbol manipulation language be used. Conventional languages are number manipulation languages; they are good at number crunching jobs. However, they are difficult to use in building expert system applications.

a. LISP Language

The most popular AI language is LISP. LISP stands for *LIS*t *P*rocessing. There are many LISP dialects; the most widely used dialects are

MACLISP and INTERLISP [Wilensky 1984]. In order to standardize LISP, in recent years COMMON LISP [Steele 84] is emerging as the standard.

One special feature of LISP is that it treats data and the program in the same way. In a conventional language the program can only manipulate data. However, a LISP program can treat any segment of program as data, and thus is able to modify it. This capability enables it to modify its knowledge base.

The smallest element in LISP is called an atom. An atom can be a variable name, a symbol, or a number. Putting a few atoms between parentheses, we have a list. An internal linked list structure stores the list. The list is also called a *s-expression* (symbolic expression). The first list following has two atoms 'diameter' and '1.0'. The second list binds the result of diameter minus .01 to diameter.

```
(diameter 1.0)
(setq diameter (diff diameter .01))
```

An atom or a list can be bound to a variable using **setq** as well. The list can then be manipulated to retrieve any of the atoms.

```
(setq shape_machinable '(thru_slot step flat_surface 2.5D_surface))
```

When the next list is evaluated, atom thru_slot will be returned.

```
(car shape_machinable)
```

On the other hand function **cdr** will return a list containing the rest of the original list, that is (step flat_surface 2.5D_surface).

There are also many functions available in LISP to manipulate data structures, lists, symbols, vectors, arrays, hunks, and structures, perform arithmetic calculations, handle input/outputs, interface with operating system functions, interface with foreign functions, define new functions, help debugging, and control the execution flow.

User functions can be defined in the same manner.

```
(defun end_cl_angle (radius)
  (cond
    ((<= radius 0.09375) '(20 28))
    ((and (<= radius 0.1875) (> radius 0.09375)) '(17 20))
    ((> radius 0.5) '(9 17))))
```

The above statement defines a function called end_cl_angle. It selects an end clearance angle for the given tool. The parameter of the function—'radius'—is the tool radius. The three s-expression following **'cond'** will be evaluated. Each s-expression in the **cond** clause has the effect of an "if ... then ..." statement. A range of end clearance angles is returned if the radius satisfies the condition.

b. PROLOG Language

PROLOG is a logic-based language [Sterling and Shapiro 1986]. A built-in inference engine uses a backward chaining control scheme to find the truth. PROLOG also offers sophisticated pattern matching functions and relational database facilities. The advantage of having a built in inference engine can often be the disadvantage. Since the user has limited control of program operation, it is easy to get into a combinatorial explosion during the search. Fewer expert systems are written in PROLOG than LISP. An expert process planning system, SIPP [Nau 1985], uses PROLOG as the base language. A frame system is built on top of PROLOG. However, the inferencing is not controlled by PROLOG but by the frame system.

PROLOG uses predicate calculus as its bases. A predicate has the form $A(x_i, i = 1,\ldots,n)$. Where A is the name and x_i s are arguments. For example, type(hole, hole_1), hole is a predicate with two arguments. The arguments can also be variables, e.g., type(hole,H). In PROLOG, words written begin with upper-case character stand for variable. When the second statement is entered, PROLOG tries to substantiate the variable 'H'. After 'H' has been substantiated, the statement becomes a sentence in propositional calculus.

c. OPS5

OPS-5 [Brownston, et al. 1985] which is commercially available from Digital Equipment Corporation, is a rule-based production system. A rule-based production system uses data-sensitive unordered rules as the basic unit of computation. A rule can also be called production rule or just production. A production system typically consist of three major parts: working memory (data), rules, and inference engine.

It has a built in forward chaining control. OPS-5 has been used in many successful commercial expert systems. OPS-5 is built on Lisp and is available on many UNIX and VMS based computers.

5.7.2 Expert System Shells and Building Tools

Since writing a complete expert system using one of the programming languages requires a tremendous amount of work, it is a good idea to try an expert system shell or a building tool. An expert system shell is an expert system stripped of its knowledge. What is left are the inference engine and support facilities. A few well-known expert system shells came from some early successful expert systems. For example, KAS came from PROSPECTOR, a mineral exploration expert system [Duda, et al. 1978]. EMYCIN is the skeleton of MYCIN [Shortliffe 1976], a bacterial infection diagnostic and treatment expert system. The TI Personal Consultant is an EMYCIN implementation on personal computer. Using expert system shells for expert process planning development can speed up the development time. However, shells were developed for other applications; the process planning application must be fitted into the existing shell. Often compromises are made. It will also be difficult to interface the shell with the design system or other process planning sub-functions. The shells may be used for quick prototyping.

Although often expert system building tools are also called expert system shells by many users, they are different. Expert system building tools provide tools for users to build their own application. They provide a language for writing rules or frames or both. KEE (Knowledge Engineering Environment) is a full-fledged expert system building tool from Intellicorp. KEE provides an excellent development environment, complete with editor, graphics interface package, explanation facility, graphics display showing rule chaining, foreign function interface, and database hooks. Multiple knowledge bases are allowed which help keep the modular design of the system. A KEE user may use frame, rule, procedure-oriented, and object-oriented representation. Both forward and backward chaining are supported. With all this flexibility, a sophisticated protyotype system can be built. KEE is written in Common LISP. It is available in many engineering workstations. The expert process planning system QTC to be introducted in Chapter 6 uses KEE.

KnowledgeCraft (KC) is another powerful expert system building tool. It was developed and marketed by Carnegie Group and Digital Equipment Corporation. KC is very similar to KEE. It provides a similar kind of development environment. Both frame (called schema in KC) and rule (CRL-OPS a superset of OPS5) representation are supported. KC is available on DEC computers. A process planning system for electronic assembly SAPIENT [Irizarry-Lopez and Chang 1988] used KC.

Many expert system building languages and tools are also available. A good survey can be found in Waterman [1986]. Although most of them can be used for a small and less sophisticated research expert system, to use them for a serious R&D project the following points should be considered.

- Is the system efficient, both on speed and memory requirement?
- Does the system provide adequate development and run time environment?
- Does the system have an open architecture which allow us to interface it with other process planning functions, database, and functions written in other languages?
- Is it available on the hardware and OS we are using or planning to use?

- Is there adequate training and customer support?
- What is the cost of the development system including all options we need and what is the cost of the run time, if there is one?

Although it is very easy to use an expert system shell or building tool to build a five to ten rule prototype process planning system, to build a real process planning will be extremely difficult. Before any code can be written, it is essential that proper training on the implementation software is given. A thorough understanding of the software system, its capabilities, and limitations, will avoid the many problems in the future.

5.8 Conclusions

In this chapter we discussed the basic structure of an expert system, the strategies used in process planning, inference engine, and knowledge representation. Readers should be reminded that although the general principles of process planning stays the same, the designs of an expert process planning system can be many. The choice of representation depends on the need. Although an ideal expert process planning system should have every function discussed in this book, a realistic system may include only a small subset of it. For example, for the CAD interface, automated feature recognition is ideal. However, since the technology is not mature enough to be robust, in practice something less sophisticated may be used. A human directed CAD interface, such as the one used in TIPPS [Chang and Wysk 1985] may be a practical solution in an industrial system. Of course, when the problem domain is more restricted, sophisticated methods may prove to be useful. For building the research and the future system, we always need to strive for the best one with the highest sophistication. Several examples have been given in the chapter to illustrate how a system can be built. However, the readers are left to decide exactly what is most appropriate for his or her application. For a more detailed example, readers are referred to the next chapter.

References

Brownston, L., Farrell, R., Kant, E., and Martin, N., *Programming Expert Systems in OPS5, An Introduction to Rule-Based Programming*, Addison-Wesley, 1985.

Buchanan, B.G., Barstow, D., Bechtal, R., Bennett, J., Clancey, W., Kulikowski, C., Mitchell, T., and Waterman, D., "Constructing an

Expert System," in *Building Expert Systems*, (eds. Hayes-Roth, F., Waterman, D.A., and Lenat, D.S.), Addison-Wesley, 1983.

Chang, T. C., Anderson, D. C., and Mitchell, O. R., "QTC—An integrated design/manufacturing/inspection system for prismatic parts," Proceedings of the ASME, 1988, *Computers in Engineering* Conference, San Francisco, Calif., vol 1, and July 31–August 3, 1988, pp. 417–426.

Chang, T.C. and R.A. Wysk, *An Introduction to Automated Process Planning Systems*, Prentice Hall, Englewood Cliffs, N.J., 1985.

Chang, T.C., "TIPPS: A totally integrated process planning system," Ph.D. Thesis, Virginia Polytechnic Institute and State University, Blacksburg, Va., 1982.

Cutkosky, M.R., Tenenbaum, J.M., Muller D., "Features in process-based design," Proceedings of the 1988 ASME International Computers in Engineering Conference and Exhibition, ASME, July 31–August 4, 1988, pp. 557–562.

Descotte, Y. and Latombe, J.C., "GARI: A problem solver that plans to machine mechanical parts," in Proceedings of IJCAI-7, 1981, pp. 766–772.

Duda, R., Hart, P.E., Nilsson, N.J., Barrett, P., Gaschnig, J.G., Konolige, K., Reboh, R., and Slocum, J., "Development of the PROSPECTOR consultation system for Mineral Exploration," SRI Report, Stanford Research Institute, October 1978.

Dunn, M.S., and Mann, W.S., "Computerized production process planning," Proceedings of the 15th NCS Annual Meeting and Technical Conference, 1978.

Giusti, F., Santochi, M., and Dini, G. "COATS: An expert system for optimal tool selection," *Annals of CIRP*, vol. 35/1/86, 1986, pp. 337–340.

Hayes, C. C., Wright, P. K., "Setup planning in machining: An expert system approach", Procs. of the 1989 NSF Conference on Advances in Manufacturing System Integration and Process, Berkeley, Calif., January, 1989, pp. 441–443.

Henderson, M.R., and D.C. Anderson, "Computer recognition and extraction of form features: A CAD/CAM link," *Computers in Industry*, vol. 5, 1984, pp. 329–339.

Henderson, M.R., "Extraction of feature information from three dimensional CAD data," Ph.D. Thesis, Purdue University, West Lafayette, Ind., 1984.

Hummel, K.E. and Brooks, S.L., "Symbolic representation of manufacturing features for an automated process planning system," in Bound Volume of the Symposium on Knowledge Based Expert Systems for Manufacturing, pp. 233-243, the Winter Annual Meeting of the ASME, Anaheim Calif., December 7–12, 1986.

Inui, M., Suzuki, H., Kimura, F. and Sata, T., "Extending process planning capabilities with dynamic manipulation of product models," Proceedings, 19th CIRP International Seminar on Manufacturing Systems, Penn State University, June 1–2, 1987, pp. 273–280.

Irizarry-Lopez, V.M., Chang, T.C., "Knowledge based process planning for electronic assembly," Second International Symposium Robotics and Manufacturing Research, Education and Applications, Albuquerque, N. Mex., November 16–18, 1988.

Iwata, K., Kakino, Y., Oba, F., and Sugimura, N., "Development of non-part family type computer aided production planning system CIMS/PRO," in Advanced Manufacturing Technology, ed. P. Blake, pp. 171–184, North-Holland Publishing Company, Amsterdam, 1980.

Joshi, S., Vissa, N.N., and Chang, T.C., "Expert process planning system with solid model interface," *International Journal of Production Research,* vol. 26, no. 5, 1988, pp. 863–885.

Kakino, Y., Ohba, F., Moriwaki, T., and Iwata, K., "A new method of parts description for computer aided process planning," in Advances in Computer-Aided Manufacturing, ed. D. McPherson, pp. 197–213, North-Holland Publishing Co., Amsterdam, 1977.

Kanumury, M., and T.C. Chang, "Survey of process planning systems for turned parts," Engineering Research Center on Intelligent Manufacturing Systems, Purdue University, January, 1987.

Kramer, T.R. and Jun, J.-S., "Software for an automated machining workstation," Report, National Bureau of Standards, July, 1986.

Matsushima, K., Okada, N., and Sata, T., "The integration of CAD and CAM by application of artificial intelligence techniques," *Annals of CIRP,* vol. 31/1/82, 1982.

Mouleeswaran, C.B., "PROPLAN: A knowledge based expert system for manufacturing process planning," Master's Thesis, University of Illinois at Chicago, 1984.

Nau, D.S. and Gray, M., "SIPS: An approach of hierarchical knowledge clustering to process planning," Bound Volume, The ASME Winter Annual Meeting, Anaheim, Calif., December 1986.

Nau, D.S., "Expert Computer Systems," *Computers,* February 1983.

Nau, D.S., *SIPP Reference Manual,* Technical Report 1515, Computer Science Department, University of Maryland, June 1985.

Shortliffe, E.H., *Computer-based Medical Consultations: MYCIN,* Elsevier, New York, 1976.

Steele, G.L., Jr., *COMMON LISP, The Language,* Digital Press, 1984.

Sterling, L. and Shapiro, E., *The ART of Prologced Programming Techniques,* The MIT Press, 1986.

Tsang, J.P., "The propel process planner," Proceedings of the 19th CIRP International Seminar on Manufacturing Systems, Pennsylvania State Univerrsity, 1987, pp. 71–77.

Tsatsoulis, C. and Kashyap, R.L., "A System for knowledge-based process planning," *Artificial Intelligence in Engineering,* vol. 3, no. 2, 1988, pp. 61–75.

Unger, M.B., and Ray, S.R., "Feature-based process planning in the AMRF," Proceedings of the 1988 ASME International Computers in Engineering Conference and Exhibition, ASME, July 31–August 4, 1988, pp. 563–569.

van 't Erve, A.H., and Kals, H.J.J., "XPLANE, a generative computer aided process planning system for part manufacturing," *Annals of the CIRP,* vol. 35, 1986, pp. 325–329.

van 't Erve, A.H., "Generative computer aided process planning for part manufacturing, an expert system approach, Ph.D. Thesis, Twente University of Technology, Netherlands, 1988.

Vissa, N., "A frame-based generative process planning system for machining prismatic parts," unpublished MS Thesis, Purdue University, 1987.

Wang, H-P. and Wysk, R.A., "An expert system for machining data selection," *Computers and Industrial Engineering,* vol. 10, no. 2, 1986, pp. 99–107.

Waterman, D.A., *A Guide to Expert Systems,* Addison-Wesley, 1986.

Wilensky, R., *LISPcraft,* Norton, 1984.

Wysk, R.A., "An automated process planning and selection program: APPAS," Ph.D. Thesis, Purdue University, West Lafayette, Ind., 1977.

Wysk, R.A., T.C. Chang, and I. Ham, "Automated process planning systems—an overview of ten years of activity," Proceedings, First CIRP Working Seminar on Computer Aided Process Planning, Paris, France, January 22–23, 1985.

6

QTC—An Example Expert Process Planning System

In previous chapters we have discussed design representation, design interface, process knowledge representation, and expert system formulation. In this chapter, an expert process planning system developed by the author will be discussed. It will be used as an example to illustrate how the ideas mentioned previously can be put together. The concept of the system design, the architecture of the system, and the design details will be addressed. The overall system discussed here is called QTC, which stands for Quick Turnaround Cell. The QTC system is capable of not only doing process planning, but also design, cell control and vision inspection. Although we will explain the overall system, the emphasis of this discussion is on the process planning function of the system.

The environment in which QTC is designed is for one-of-a-kind prismatic parts which requires immediate attention. The part handled by the system usually can be machined by a machining center. The system is especially useful in a research and development environment where quick part turnaround means shorter development time. In this environment the demand on process planning is extremely high. Since every part that comes into the system is different, new plans and part programs need to be generated constantly. It is, therefore, desirable to automate as much of the planning decision making as possible. The QTC system is intended for one-of-a-kind part production; it rules out the possibility of using part family concepts for process planning. It must generate new plans based on the knowledge of basic machining processes, tools, and fixturing methods. In order to produce the part quickly, the optimization of the process plan and cutter path becomes less important than the use of available methods and tooling. It is also desirable to integrate all sub-modules of the system into a coherent unit. The design database should be shared by the designer, the process planner, the NC operator, and the vision inspection system.

Based on the above discussion, the requirements of the system can be summarized as:

- One-of-a-kind prismatic part machining
- Utilizes the available resource to produce the part quickly
- Integrated design/process planning/inspection system
- Other than design, no human decision making is necessary

The above mentioned requirements are primary requirements. Based on the primary requirements, there are also secondary requirements. The secondary requirements are derived from the primary requirements. They can be as important as the primary requirements for the successful implementation of the system. The following is a list of the secondary requirements:

- User friendly design environment
- No delay on stock preparation
- Minimum tooling preparation
- Detailed process plan generation
- Automated part program generation
- Adapting to the cell capability change
- Effective cell control and monitoring
- Automated inspection

A user friendly design environment is necessary in order to simplify the design. This design environment should interface with a solid modeller, thus a complete and unambiguous design model can be generated for process planning use. Technological information such as tolerances and surface properties should also be part of the model and are easily entered through the design system. In a real production environment stock preparation can be a lengthy process. For the quick turnaround purpose, delay of stock preparation should also reduced to a minimum. The tooling preparation is the same as the stock preparation; the potential delay should be eliminated. A detailed process plan and finished part program can further reduce the possible human delay. Therefore, the automatic generation of such information is necessary. Further, the plans generated are to be executed soon, and the plan should take advantage of the current cell status and resources. The machining cell must be properly controlled by the system, and its operation should be monitored. Finally, in order to ensure the parts produced satisfy the design specification, they should be inspected quickly. To quickly inspect one-of-a-kind parts, it is necessary to have automated this process.

In order to achieve the above requirements, the QTC research group proposed the following approaches. Figure 6.1 shows the operation sequence of the system concept.

- Use pre-cut raw stock
- Feature-based design interface
- Dual model representation (both feature and boundary representation)
- Feature refinement to obtain the final manufacturing features
- Intelligent process planning utilizing system status data
- Standard process plan documentation format
- Embedded solid modeler provides geometry evaluation and manipulation for the entire system
- Cell controller provides user interface between the machine and the operator
- Cell monitoring (including tooling and setup monitoring) by vision
- Finished part inspection by vision
- Existing available raw stock, tools, and processes are used.

The rationale behind using pre-cut raw stock is to minimize the time required for raw stock preparation. Since the QTC system is designed to handle prismatic parts, the raw stock used is box shape stock. Stock data are stored in the system and used by the designer as the building block for designing new parts. A database interface should provide the design with in stock, raw stock information.

The design system has a feature-based construction. The features correspond to the volumetric features machinable by the machine tools in the cell. In order not to burden the designer with manufacturing details, the design sequence has no relationship to the machin-

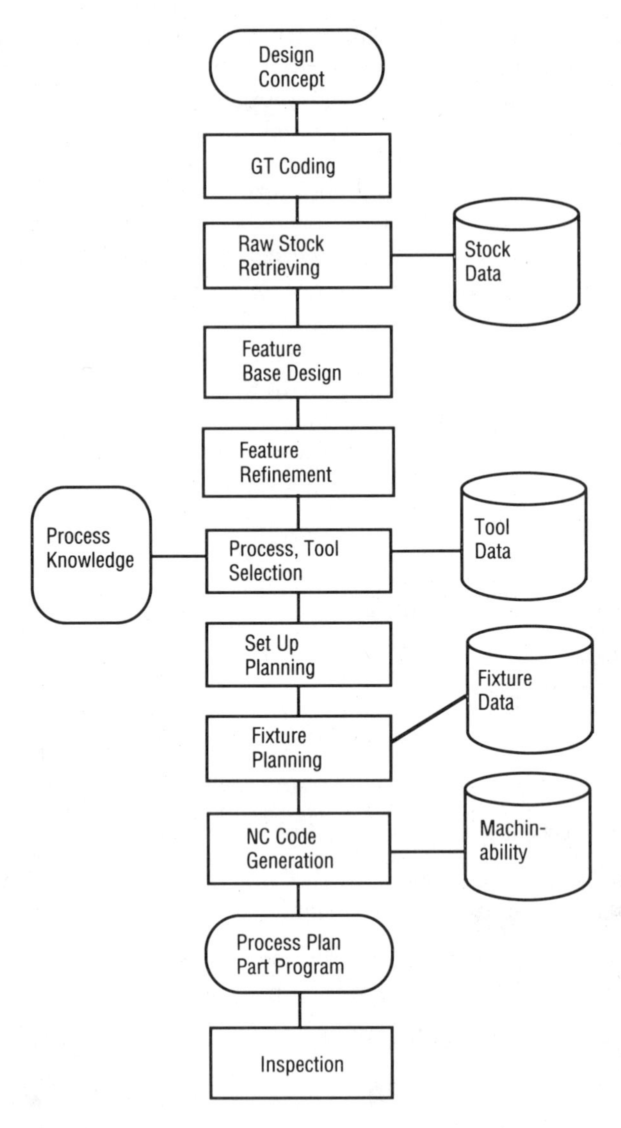

Figure 6.1 QTC operation flow diagram

ing sequence. A 3D solid modeller is linked to the design system to provide drawings of the design. The feature-based design system is independent of the solid modeller. Therefore, it allows for future expansion of the system to include other, more powerful solid modellers. After a design is completed, the design data—a feature file—is sent to the process planning system. The process planning system needs to reason about the exact manufacturing features, their relationships, and feasible approach directions. This information is then sent to an expert process planner for processes, tools, fixture selection, and process sequencing. This geometric reasoning procedure is called feature refinement. For the feature refinement and NC cutter path generation, a solid modeller is needed to evaluate the model. This solid modeller can be a different one than the one used in the design system. In this project, an in-house built polygonal modeller called TWIN [Mashburn 1987] is used.

A standard process plan documentation format is needed for passing the process plan information to the cell controller. The process plan documentation format should be designed in such a way that it is complete, unambiguous, and easy to interpret. Pointers in the process plan documentation are used to link the cutter path files. The cell controller must be able to interpret the process plan documentation and post process the cutter path files into machine specific part program. The cell controller should also provide an interface to the data bases. It must be easy to use and require a minimum amount of typing.

Finally, an inspection system is needed for tooling, setup, and finished part monitoring and inspection. Ideally, the inspection results should be used by the process planning to update the process knowledge. However, due to the difficulty in machine learning, the QTC system does not consider this inspection feedback.

6.1 The QTC System Architecture

The QTC system consists of four major functional modules: design, process planning, cell control, and inspection (Fig. 6.2). A more detailed system architecture is shown in Fig. 6.3. Data files are used

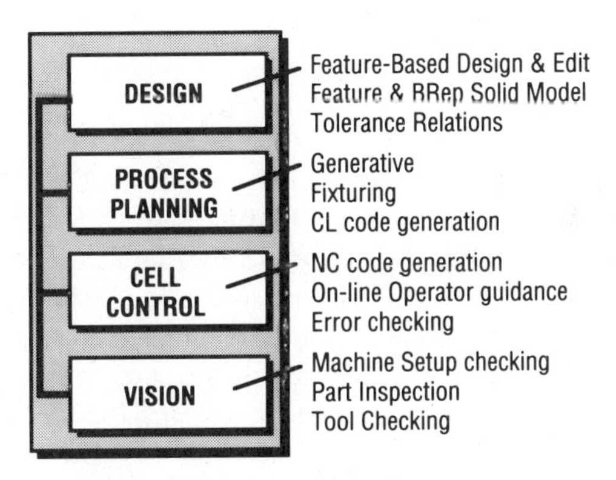

Figure 6.2 QTC system overall structure

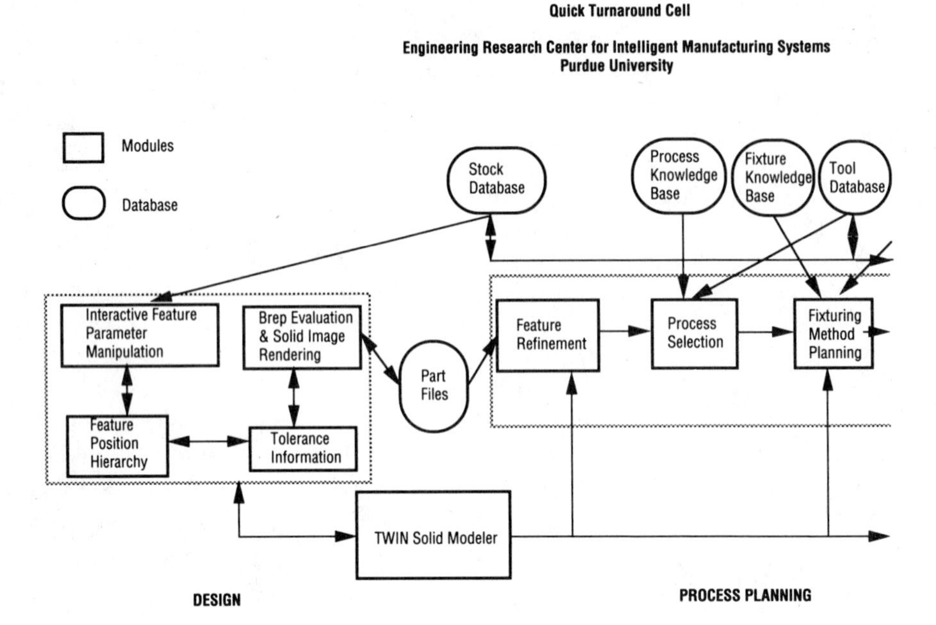

Figure 6.3 Detailed QTC system architecture

as the interface between functional parts. This arrangement allows each functional part to be developed incrementally, without having a drastic effect on the overall system structure. There are two major human interfaces within the system. The first one is naturally the interface with the designer. Process planning is done automatically, with no human interface necessary. The cell control module has an interactive graphical user interface to facilitate detailed machine operator instructions.

6.2 The QTC Design System

The structure of the design system is shown in Fig. 6.4. As mentioned, the design environment is feature-based, where each feature is loosely related to machining operations such as a hole, slot, or counterbore. These features are not stringent manufacturing features. For example, a slot can carry a lot of meanings. It can be blind, through, or even degenerate into a pocket (Fig. 6.5). The designer is free to use the feature as he or she wishes. The feature refinement module in the process planner analyzes the design model and further classifies designed features into manufacturing features. The features initially represent material removal operations, only a difference

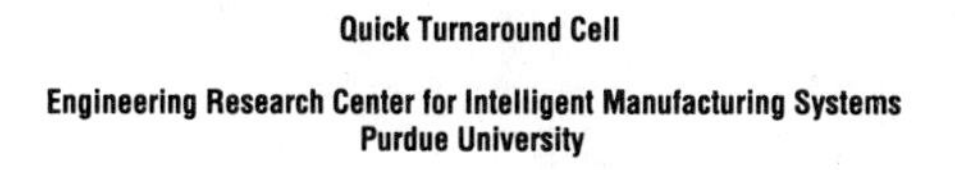

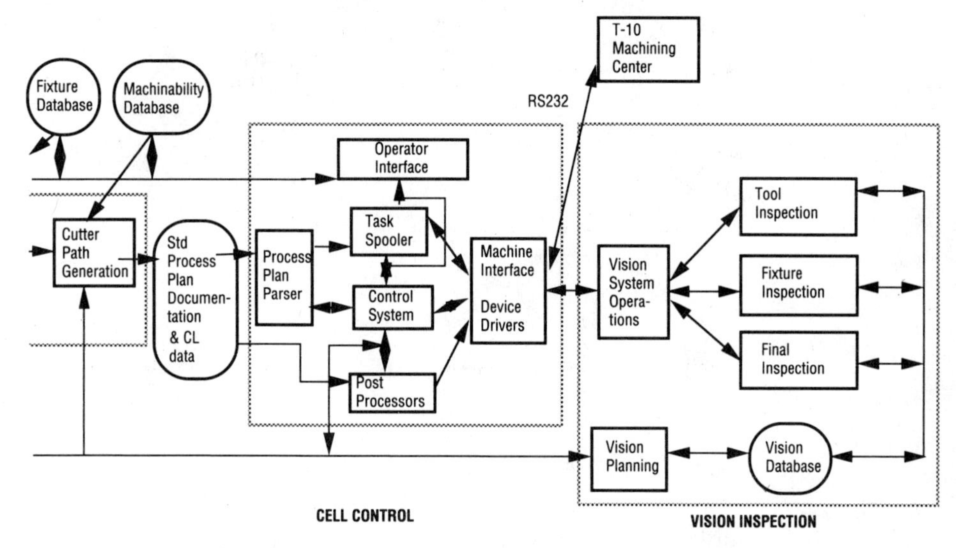

Figure 6.3 (continued)

operator is used in the design system. Therefore, each feature corresponds to a cavity volume to be removed by machining. This simplifies the planning greatly; however, the flexibility of design is limited. In the second phase of QTC, a union operator is included. Feature recognition techniques, such as the ones discussed in Chapter 3, are incorporated into the system.

The design module and design model data have been implemented using an object-oriented approach for dealing with features and their components. While all features have common physical attributes, such as position or orientation, they also have common actions they must perform. For example, they must be drawn or have their parameters interactively changed. By using an object-oriented approach, different features can be dealt with similarly. This isolates the "local" characteristics of the individual features from the "general" functioning of the system and facilitates the addition of new features.

To assist the designer in interacting with the three dimensional geometry of the features and in the construction of the model, graphical entities called handles are used. Handles are characteristic geometric elements of features, representing points and lines of interest. For example, point handles are used to represent the vertices of a rectangular workpiece or the endpoints of the axis of a hole,

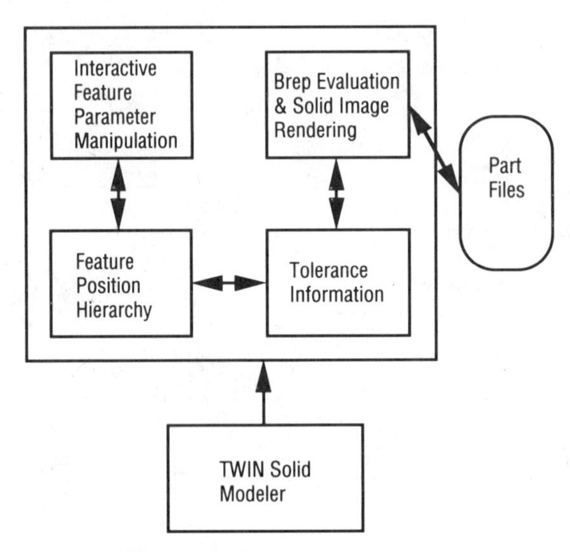

Figure 6.4 The design system

while line handles are used to represent geometric parameters, such as the length of a slot or the depth of a pocket (Fig. 6.6). By interacting with a feature through its handles, the designer can easily position or orient the feature, or change the values of its geometric parameters.

In order to provide enough information for process planning to make the part, tolerance information is included in the feature model.

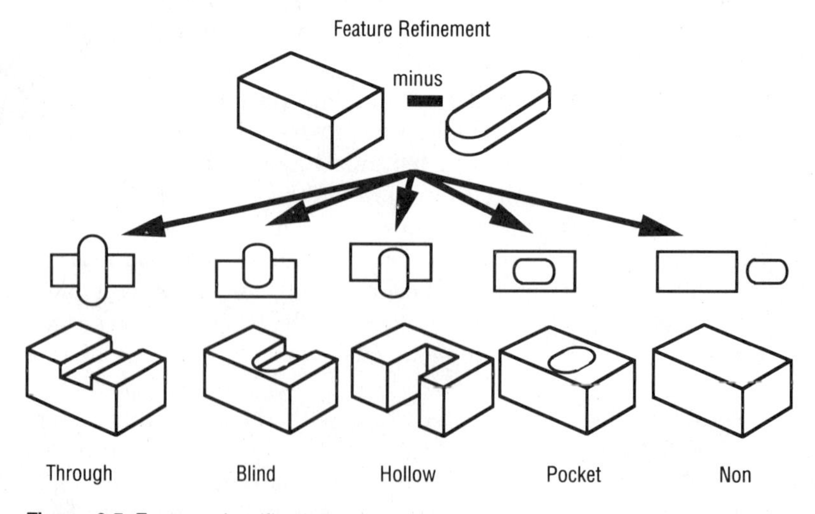

Figure 6.5 Feature classification

Feature

Common Data
Handles:

Dimensions
Tolerances
Surface finish

Figure 6.6 Feature data structure

Each feature has bilateral tolerances associated with its size parameters, such as depth or radius, and tolerances on its relative position. Also, there are form tolerances on features, such as surface finish or cylindricity.

Point handles represent geometric locations of interest on the features and can be used by the designer to establish reference points on features. Corresponding to the way machining is done face-by-face, a default coordinate system (Fig. 6.7) is established on each face of the raw material where machining operations are possible. New references can be established by the designer on any point handle of a feature, on the current work face, or the workpiece which is also considered a feature. Other features can be positioned relative to this new reference. This establishes a positional relationship between the two features which is maintained in the feature model database. The absolute position of the new feature and its associated tolerance stackup can be calculated based on the relative position of the new feature and the absolute position of the reference feature.

The tolerance scheme can be represented by a position vector V_i and a tolerance matrix T_i..

$$V_i = \begin{bmatrix} DX_i \\ DY_i \\ DZ_i \end{bmatrix} \qquad T_i = \begin{bmatrix} TX_i^1 & TX_i^2 \\ TY_i^1 & TY_i^2 \\ TZ_i^1 & TZ_i^2 \end{bmatrix}$$

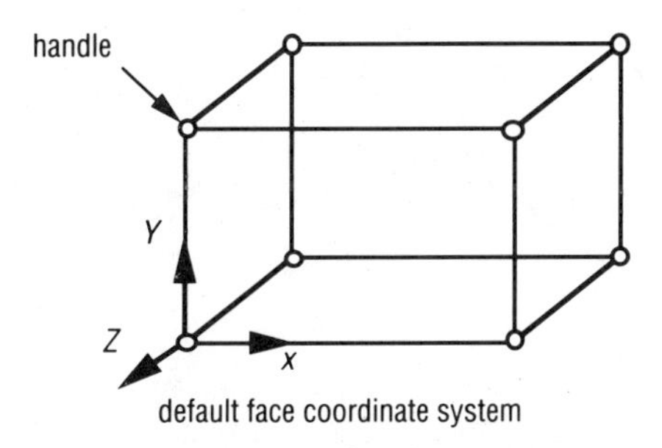

Figure 6.7 Default face coordinate system and handles

where

DX_i, DY_i, DZ_i : magnitude
TX_i^1, TY_i^1, TZ_i^1 : negative tolerance
TX_i^2, TY_i^2, TZ_i^2 : positive tolerance

When there is a chain of tolerance (Fig. 6.8), the position vector V_3 which defines the position of the last feature with respect to the reference is:

$$V_3 = V_1 + V_2$$

and the tolerance is:

$$T_3 = T_1 + T_2$$

However, if V_3 and T_3 are known, to find V_2 and, T_2 one would do:

$$V_2 = V_3 - V_1$$

$$T_2 = T_3 + T_1$$

The tolerance is always additive. By applying this stacking method, a relative position can be found. Frequently, one feature is defined with reference to a different handle on another feature. A position vector can be established between the two handles on the same feature. Then, this position vector is treated the same as the position vector between two different features.

The final representation of a part is a data structure which consists of a list of features. This data structure is written to an ASCII file for storage. The part file is used to re-create the part data structure for use by the process planning, the cell control, and the vision inspection functions.

The design system calls TWIN to produce a polygonal boundary representation of the part. During the design stage, TWIN is used to generate a boundary representation for display and verification by the designer. Two graphics packages display the picture in either wire-

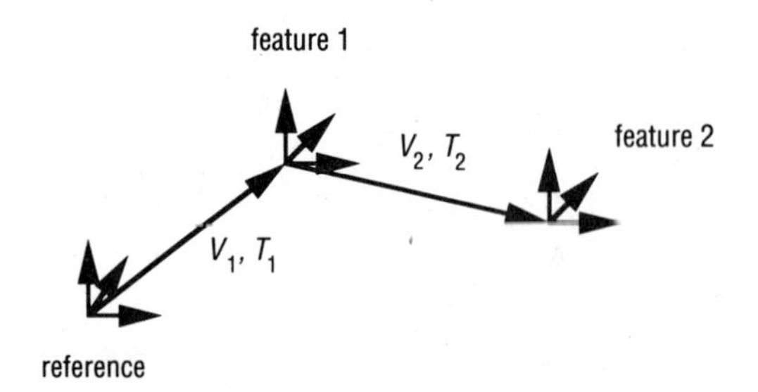

Figure 6.8 Position and tolerance chain

frame or shaded image of the evaluated boundary representation. The only data passed to the subsequent system functions is the feature model consisting of a list of features, starting with the workpiece, their parameters, and the handle-based positioning information. The boundary model is re-generated as it is needed.

6.3 Process Planning

The process planing module of the QTC system is called AMPS (Automatic Machining Planning System) [Kanumury 1988]. AMPS consists of feature refinement, process selection, tool selection, process sequencing, fixturing method planning, and NC cutter path generation—six functional modules. Each functional module is related to each other. AMPS planner is implemented in an expert system shell KEE® [Intellicorp]. Feature refinement module, tool data base management system, fixturing method planning, machinability data base management system, and NC cutter path generation modules are written in C language. These C functions are interfaced with KEE through a KEE®/C interface package. In the following, details of the function modules of process planning are discussed.

6.3.1 Feature Refinement

The first step in process planning is feature refinement. The purpose of feature refinement can be illustrated by the design example given in Section 6.2. Since the design system utilizes the design oriented manufacturing features, it is necessary to find out the above information before one starts the subsequent process planning tasks. The feature refinement problem is somewhat similar to that of the feature recognition problem, as some of the methods developed in feature recognition [Joshi and Chang 1987] are also used in feature refinement.

There are three major tasks in the feature refinement:

- feature classification
- features relation identification
- feasible approach and feed direction determination

The feature refinement module uses both the feature model and the boundary model to reason about feature interactions. A feature model is used to provide higher level information such as feature type, position reference, tolerances, and surface finish. The exact boundary of a feature in the finished part is formed using the boundary model. Feature relations, approach direction, and feed direction determination require boundary information as well.

6.3.1.1 Feature classification

The TWIN solid modeler is again used during the feature refinement stage to generate the boundary model. There are two problems associated with the approach of having two models. The first one is how to relate surfaces on a boundary model to the feature model (and vice versa). The second one is due to the fact that TWIN is a polygonal model, the boundary representation is not exact. There is not only difficulty in distinguishing curved surface features from a boundary model, but there is also difficulty in generating the NC cutter path (need exact boundary). The approach taken is to add a pointer to the TWIN data structure. Faces in a boundary are tagged with feature information. A tag is two types: face type and feature type. Face type can be cylindrical face, top flat face, bottom flat face, side face, etc. It denotes the specific face in a feature. Figure 6.9 illustrates tags on a cylindrical surface. Another example can be found in Fig. 6.10. By collecting all faces which have the same tag, a feature can be found.

The exact boundary instead of the original designed feature boundary can be found as well. In Fig. 6.11, the original slot length was 10 inches. After boundary evaluation, the detected slot length is 6 inches. Figure 6.12 shows a structure linking the feature model and the boundary model. In the figure two features are highlighted. The first feature is "raw mtl" feature. It is a block consisting of six planer faces. A forward pointer (which is implied by the id "raw mtl") links the feature with a boundary representation. The second feature is "Feature 2." "Feature 2" also links to some faces in the boundary

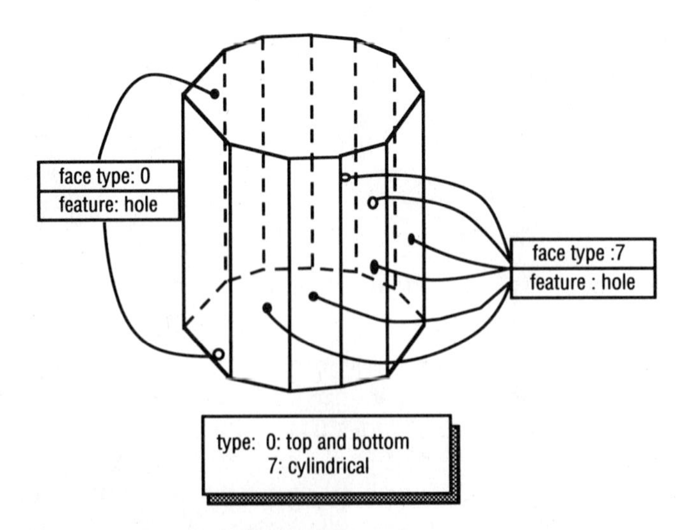

Figure 6.9 Tagging a hole feature surface

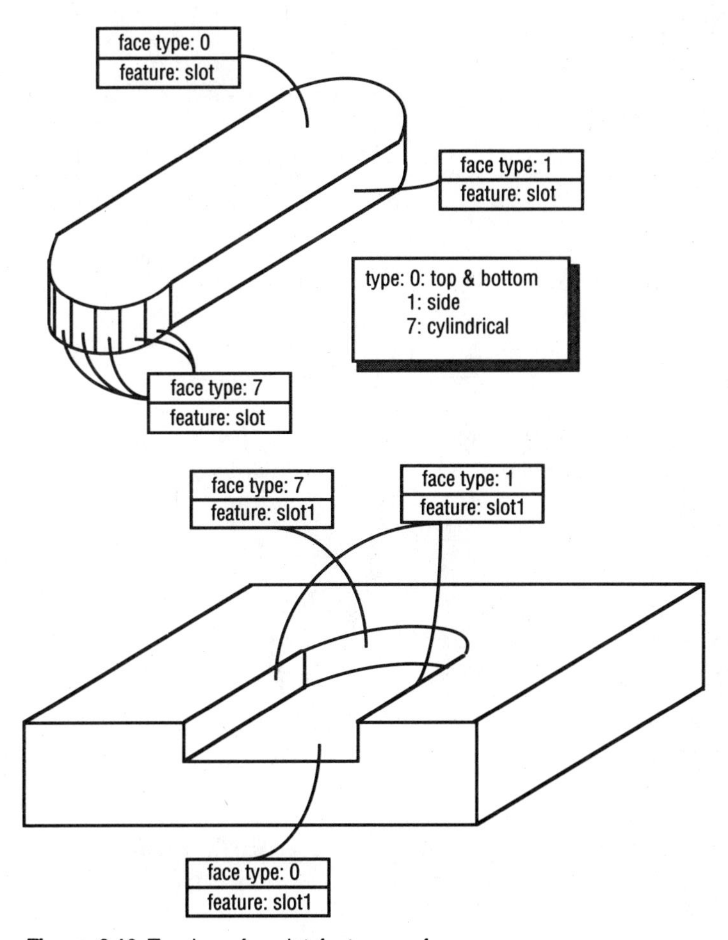

Figure 6.10 Tagging of a slot feature surface

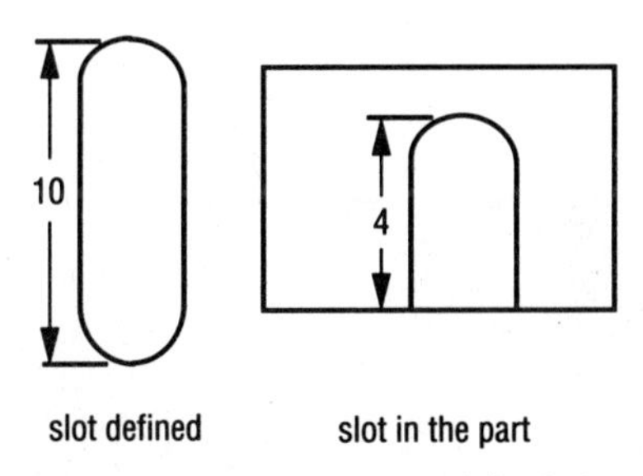

Figure 6.11 Defined slot and final slot

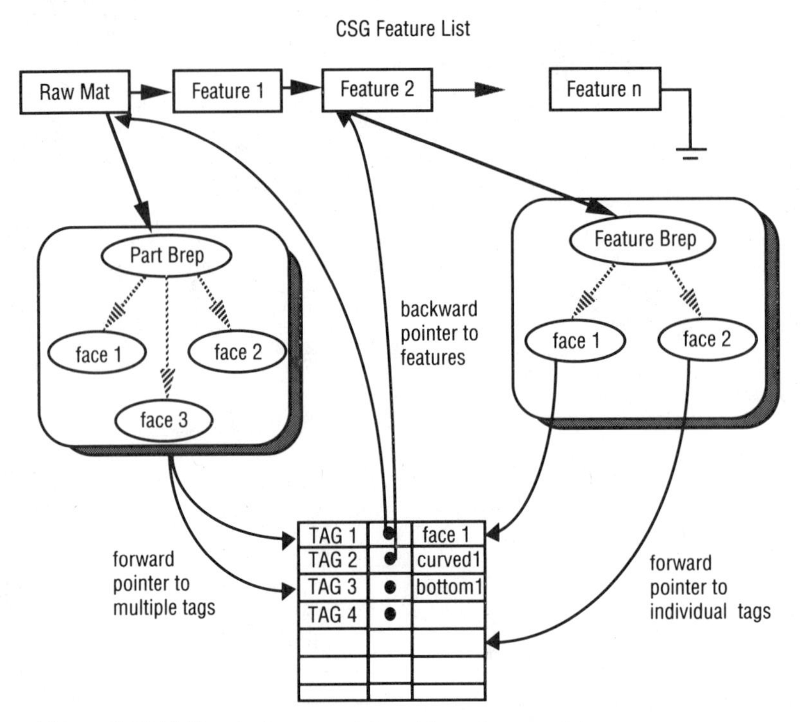

Figure 6.12 Linking features model and boundary model

representation. Backward pointers are used to link back to the feature. Through this data structure the meaning of feature can be found easily.

Features are classified into sub-classes. As shown in Fig. 6.13, three levels of feature classification are used. Implicitly the system (TWIN, or other solid modeler) works with the general solid primitives and boundary faces. The designer uses the general feature to build a part model. The process planning system needs to refine the design features into manufacturing features. The refinement is done by the following steps:

Step 1. Evaluating the feature model into a boundary model

Step 2. Grouping all faces that have the same tag into a set

Step 3. Applying feature classification rules to all face sets

Since all faces are tagged, grouping can be done easily by sorting the faces in a boundary model. IF...then rules for each sub-class can be written to do the classification. A few typical rules for slot can be expressed in English:

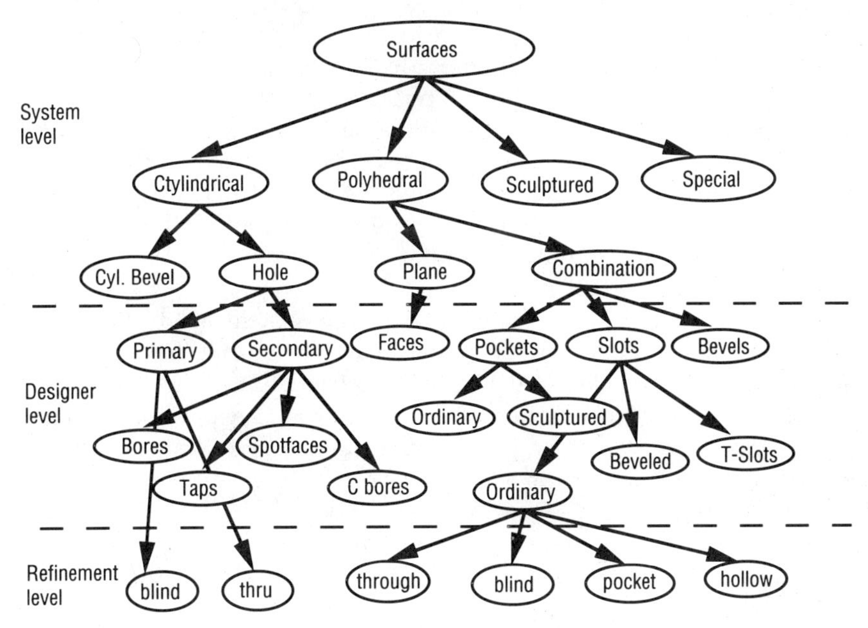

Figure 6.13 Feature classification

IF the slot feature has a missing end face and a top face THEN the slot is blind.

IF two end faces and a top face of a slot are missing THEN the slot is through.

IF both the top and the bottom faces of a slot are missing THEN the slot is hollow.

IF only the top face of a slot is missing THEN the slot is a pocket.

IF no face is missing THEN the slot is an inner cavity.

Currently, a straight forward matching algorithm is implemented in the QTC system.

There are also features which degenerate to another feature after several boolean operations. For example, a pocket may degenerate into a slot if two sides have been removed. A slot with one side and both ends missing becomes a step. In addition to the regular feature classification, the system also checks for degenerate cases.

6.3.1.2 Feature relations

As discussed in Chapter 4, the relationships between feature are important for determining the process sequence and sometimes the process itself. Features are related to one another by means of geometry and tolerances. This relationship can involve more than two features at a time. The number of possible relationships increase sharply as the number of features increase. It is not possible to take all of them into consideration. Fortunately, not every relationship is of interest to us from manufacturing point of view. Currently, the AMPS system considers only the binary geometric relationship between features. Although this greatly simplifies the problem, it is manageable and is sufficient for most cases.

What also needs to be considered in the system are nested type and intersection type relations. If two features are not nested or intersected, they are declared not to be related. Otherwise they are related in one way or another. Nested relations can be further classified as contained at bottom or contained at the side. One feature is nested in another feature if all the faces of it open up into the other feature. There are many types of intersection relations possible. The intersection between the same type of features and different types features are classified differently because of the manufacturing significance.

To find the nested relations, the boundary representation of two features under study is evaluated.

Step 1. Select any pair of features (features A and B).

Step 2. Perform a boolean union operation of the two features ($A \cup B$).

Step 3. Check the resultant boundary model for one of the following cases:

   a. a new loop is found at one of the faces in B,
   b. one of the faces in B is splitted,
   c. a new edge on the outer loop of the face of feature B is created.

Step 4. If one of the above three cases exists, then, A is contained in B.

Intersection types are coded in the system. An intersection code consists of three digits. The first two digits represent the feature types, and the third digit represents the intersection type. A different code is assigned to each of the intersecting features. For example, the contained feature and the containing feature are assigned code '1' and '0', respectively. Therefore, the relationship can be identified easily.

6.3.1.3 Merging and splitting

Merging is building a complex feature from a few simple features. Splitting separates a feature into a few new features. The purpose of

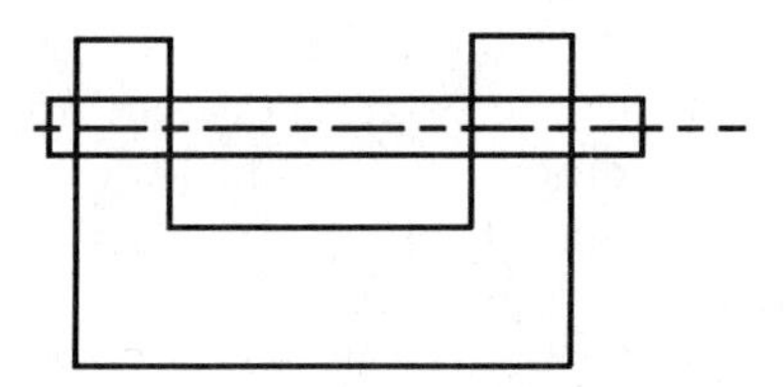

Figure 6.14 Splitting a hole into two holes

merging and splitting features is to optimize manufacturing operations. Often it is possible to machine a group of features in one setting. For example, a few connected slots or pockets can be treated as one complex pocket. A general algorithm can be developed to optimize a way to machine the generic pocket, thus simplifying the planning. Splitting is used to decompose an un-machinable feature into sub-features so they can be machined separately. Figure 6.14 shows a hole which is too long to be drilled (length/diameter ratio greater than 7) in one setup. Depending on tolerance specification, it may be drilled separately from two ends. Such a hole is then splitted into two holes of shorter depth.

The major problem of merging is the difficulty in determining to what degree of complexity a merged feature should be. When features of different types are allowed to be merged, the merged feature can be extremely complex. In this situation, the benefit of merging is lost because there are no methods yet developed for the merged feature. The principle for merging is that there be a well defined method available for the merged feature. Splitting is done only when possible manufacturing alternatives warrant it. In the AMPS system, some heuristics have been developed to handle the merging and splitting operations.

6.3.1.4 Approach and feed directions

Approach and feed directions are important for determining setups and cutter path. We first will define a few terms used in this section. The approach direction of a feature is an unobstructed path that a tool can take to access the feature in the workpiece. The feed direction of a feature is the direction of tool motion necessary to cut the feature. The through direction is an unobstructed path from the feature to the surface of the workpiece. The view direction of a feature is the direction in which the feature edges are visible but cannot be manufactured. The view direction is useful to the fixturing [Shah 1988] and the vision inspection [Park 1988]. We are interested in finding all the above directions.

In order to determine approach directions, a boundary model for the part is generated. To find the approach direction, one must check whether there is any obstacle blocking the tool approach to a feature. A simulated tool sweep volume is used to probe the features from several pre-defined directions. These pre-defined directions are the most likely approaches for a given feature. For example, for hole, the directions are positive and negative direction along the axis. Following is a procedure for approach determination.

Step 1. Evaluate the feature model of the part into a boundary model.

Step 2. For each feature do the step 3.

Step 3. Use a tool (cylinder for hole drilling, rectangular block for slot milling) to probe the feature from all pre-defined directions. (Clash the tool with the part design boundary model. When probed from the top, the bottom of the tool touches the bottom face of the feature, and the top extends above the top of the feature.)

Step 4. Those directions which do not cause interference between the tool and the workpiece are collected in an approach direction set.

The feed directions are pre-defined with the feature primitives. Coordinate transformations used on the features should be applied to these feed directions in order to obtain the final feed directions. The through direction is a special case of feed direction; it is especially significant for slots and pockets. A through slot is said to have two through directions, one from each opening. A blind slot has only one through direction. The through direction can eliminate tool plunging during the milling operation. When there is no through direction, a milling tool must plunge down from the approach direction before it can take a feed direction.

An example for finding directions for hole follows. It is worthwhile to point out that determining directions is feature specific. Algorithms need to be developed for different features.

For a hole with no bottom faces, a tool boundary model is generated on the axis in the positive and negative axis directions. The tool boundary model is a cylinder with a radius equal to 80% of the hole radius and a length equal to the longest diagonal of the workpiece boundary model.

Case 1. The tool boundary model in the positive axis direction does not clash with the workpiece boundary model. The approach direction is in the negative axis direction. The feed direction is in the negative axis direction. The through direction is in the positive axis direction.

Case 2. The tool boundary model in the negative axis direction does not clash with the workpiece boundary model. The approach direction is in the positive axis direction. The feed direction is in the positive axis direction. The thru direction is in the negative axis direction.

Case 3. The tool boundary model does not clash with the workpiece boundary model in either direction. The feature type is through. Otherwise, the feature type is blind.

For a hole with one bottom face, the bottom normal is determined. In determining approach and feed directions, the face normals point 'outward' from the workpiece boundary model. The approach direction is in the negative direction of the bottom normal, the feed direction is in the negative direction of the bottom normal, the thru direction is in the positive direction of the bottom normal.

Case 1. The bottom face of the hole does not contain another adjacent hole. The feature type is blind. Otherwise, the feature type is thru-one-end.

Case 2. The hole is contained within an adjacent hole. The auxiliary approach direction is determined from the approach and auxiliary approach directions of the adjacent hole.

For a hole with two bottom faces, if both bottom faces do not contain other holes, the hole cannot be machined.

6.3.1.5 Feature refinement results

The results from the feature refinement module are represented in a frame format and passed to the process selection module. For slot1 in Fig. 6.15, the sample output is shown below:

```
(slot_1
  ((id 2)
  (type 1)
  (feat_type 0)
  (length 4.50000)
  (width 0.80000)
  (depth 0.500000)
  (b_type FLAT)
  (handle_face_ref 0)
  (ref_feat work_piece)
  (ref_face face1)
  (surface_finish 250.000000)
  (length_tol_range (-0.050000 0.050000))
  (width_tol_range (-0.050000 0.050000))
```

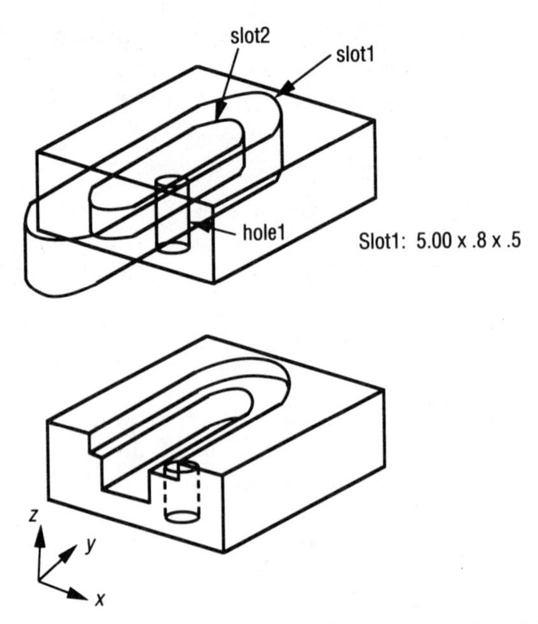

Figure 6.15 Example part design, feature model and boundary model

```
(depth_tol_range (-0.050000 0.050000))
(locational_tol_range (-0.086603 0.086603))
(parallelity_tol_ref ())
(perpendicularity_tol_ref ())
(angularity_tol_ref ())
(approach_direction ((0.000 0.000 -1.000)))
(aux_approach_direction ())
(feed_direction ((0.000 1.000 0.000)
  (0.000 -1.000 0.000)))
(thru_direction ((0.000 1.000 0.000)))
(thru_type THRU_ONE_END)
(contains_b (hole1))
(contains_s ())
(contained_b (slot1))
(contained_s ())
(merge ())
(split ())
(causing_split ()))
```

Most of the data in the frame are self-explanatory. Note that the length of the slot has been modified from the original design value. The 'contains_b' slot indicates that 'hole1' is contains at the bottom of 'slot1'. 'contined_b' slot shows that 'slot1' is contained under 'slot3'. This refined data supplies complete information for the subsequent process planning functions.

In the implementation, the features are represented in a hierarchical frame structure. The feature hierarchy is shown in Fig. 6.13. The grouping and classification of features can increase the inferential efficiency of the system. The lower level frame (child) also inherit slots from the higher level frame (parent). Slots which are inherited by the child frame are common attributes among child frames, such as attributes of tolerances, feature relations, and geometrical data. A frame is also allowed to have slots pertinent to itself.

6.3.2 Process Selection

The process selection is done by a knowledge-based approach. Process capabilities based on our previous research results and input from industry are collected as process knowledge. The rules are written in a backward planning approach. The antecedents specifies the goal condition of a feature. The process knowledge is organized in hierarchical rules (Fig. 6.16). A process taxonomy classifies the processes into hole-making-processes and non-hole-making-processes. The hole-making-processes are further classified into fine-finishing-operations, finishing-operations, hole-generation-operations, etc. The hole-generation-operations include conventional-processes, deep-hole-processes, etc. The conventional-processes includes center-drilling, core-drilling, twist-drilling, end-milling, etc. The antecedents of the higher level classes are inherited by the lower level classes. The major advantages of this approach are that it is easy to understand and it reduces the search space. The process knowledge is also better organized for maintenance. For example, a rule for a rough boring process is shown in Fig. 6.16.

```
(ROUGH_BORING
  (FINISHING_OPERATIONS (CLASSES GENERIC))
  NIL
  ()
  ((ACTION.TYPE (TOGETHER))
   (EXTERNAL.FORM
  ((IF    (TEXT (PROCESS PROGRESS))
          (?FEAT IS IN CLASS PRIMARY_HOLE_FEAT)
          (THE L_RADIUS OF ?FEAT IS ?LRA)
          (LISP (ROUGH_BORING_SATISFY_CONSTRAINTS
                ?FEAT))
          THEN
          IN.NEW.WORLD
          (LISP (ROUGH_BORING_MODIFY_RADIUS
                (- ?LRA 0.08)
```

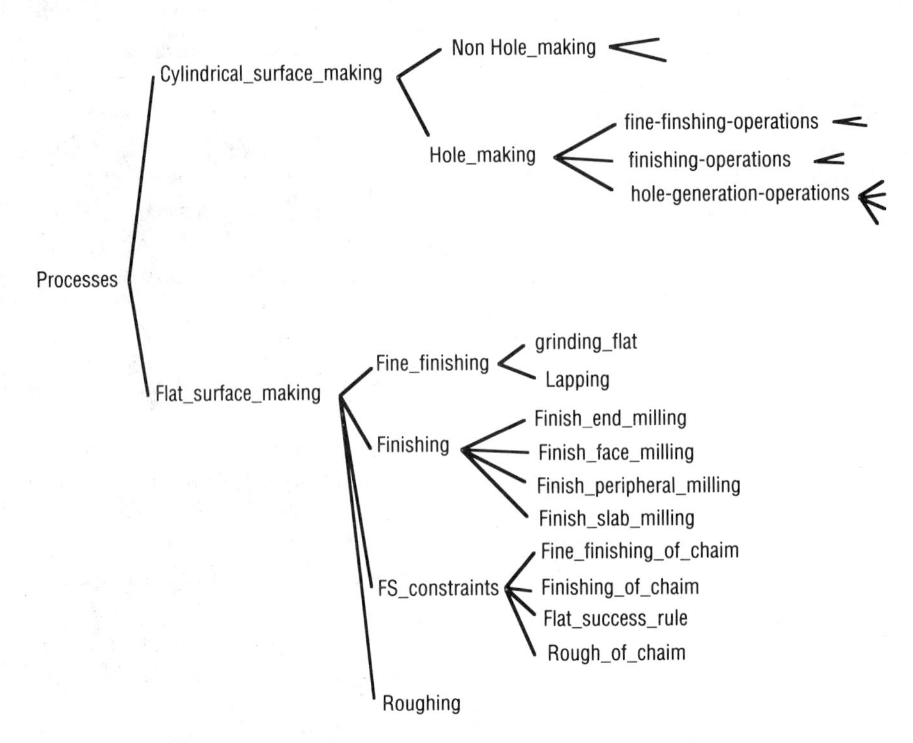

Figure 6.16 Process hierarchy

```
                  0.1
                  ?FEAT
                  $WORLD$))
         (LISP (ROUGH_BORING_MODIFY_CONSTRAINTS
                ?FEAT $WORLD$))
         (THE PROCESS OF ?FEAT IS ROUGH_BORING)
         (LISP (QUERY (PROCESS.GOAL ?FEAT)
                         'HOLE_MAKING_RULES
                  $WORLD$) ) ) ) )
   (MAKE.AND.WORLD? (NIL) )
   (PARSE.ERRORS)
   (RULE.TYPE (NEW.WORLD.ACTION)
   (WEIGHT (5.0) )
   ) )
```

This rule is written in KEE®, and some functions used in the rule are written in LISP. The variable ?FEAT refers to a frame representing the feature being planned. Two other rules are shown in Chapter 4.

6.3.3 Tool Selection

The process selection module selects a set of feasible processes for each feature. The most appropriate cutting tools are then selected.

Tool selection rules are used to synthesize tool database queries. The query returns a set of feasible tools. The tool database uses a relational schema and has an SQL [Date 1982] interface. The data base is implemented using a UNIX utility for file manipulation. The query language is developed in-house. In each data record, information on tool number, tool material, tool geometry, dimension, tool life left, and operation types are stored.

Tools are classified into three groups (Fig. 6.17): those which are currently in the tool magazine, those which have already been set up, and those which are available in the tool crib but have not been set up yet. When tools are selected the priorities are given according to the sequence in which they are mentioned above. The rationale is that the least tool preparation time should be spent in order to minimize the lead time for the part manufacturing. If an exact tool cannot be found, a tool second on the list is selected. The system also tries to use the same milling cutter for several features. Although one cutter may not be the best for all of the features, reducing the number of tools needed for the job reduces the number of tool changes. In machine shops, if only optimum cutters are used, hundreds of cutters need to be purchased and maintained. This represents a large amount of capital investment and tremendous set up and storage management problems. Also because the tool magazine can hold only a limited number of tools at any given time, the loading and unloading of additional tools creates undesirable idle time.

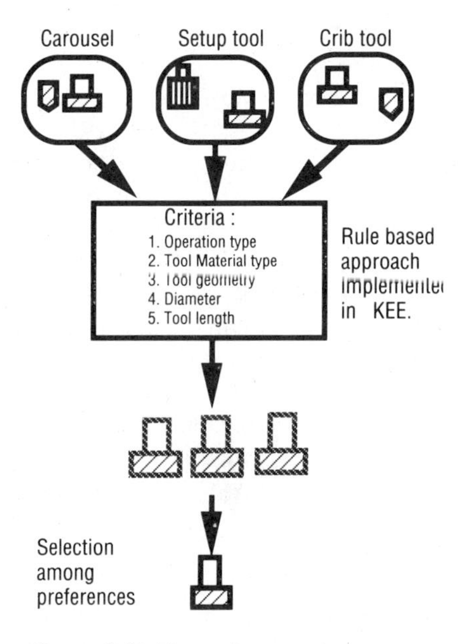

Figure 6.17 Three classes of tools

Actually tool selection involves the determination of the following:

- tool material—HSS, carbide, etc.
- tool geometry—helix angle, rack angle, etc.
- tool type—drill, mill, boring bar, etc.
- tool dimension—overall length, flute length, diameter

The tool material selection is based on the raw stock material and its hardness. Suggested tool material data are taken from Tool and Manufacturing Engineering Handbook [Drozda 1983]. Tool geometry is selected based on the feature geometry, raw material condition, and tool material. The tool type is determined by the process type. Overall tool length is the length which ensures the spindle has a collision free movement. It is determined using the following procedure:

Step 1. Find the intersection of the intermediate workpiece boundary model with a cylinder which has the same diameter as the spindle, and align the axis of the cylinder with the feature.

Step 2. Do step 1 for the final part boundary model with the cylinder.

Step 3. The minimum overall tool length is the difference between the extreme point in the spindle approach direction from step 1 and the furthest point in the spindle approach direction from step 2.

The flute length is the actual height of the refined feature. The height is determined by the approach direction. The tool diameter is determined by the refined feature dimension. For hole features the diameter is easy to determine. For slots, the width of the feature is the upper bound of the tool diameter. Although it is desirable to use the width as the tool diameter, in practice such a tool may not be available. Therefore, the diameter is the largest available tool which is smaller than the width of the slot.

Rules have been implemented to generate tool selection criterion. These rules are divided into three categories: material selection, geometry selection, and cutter type selection. A forward chaining mechanism is used in a tool selection knowledge base. The result of the tool selection rules is a query which is sent to the tool database management system. The selected tool is returned.

An example of a tool selection rule is shown below.

```
((if (the tool_selection_mode of ?feat is select)
  (the tool_frame of ?feat is ?tf)
  (?tf is in class twist_drilling_tools)
```

```
    (the material of work_piece is ?ma)
    (lisp (soft_material_drill ?ma))
    (the l_d_ratio of ?feat is ?ld)
    (lisp (?ld 3.0))
  then
    (the helix_angle of ?tf is '(35 40)))))
```

A query generated by the system is shown below.

```
(Selection: Current_carousel_no
  cost_per_tool
  tool_cumulative_hours
  estimated_tool_life_hours
  carousel_qty
  crib_qty
  inventory_qty
  diameter)
(where : generic_type = end_mill and
  tool_length >= 0.5 and
  reserved_flute_length >= 0.25 and
  diameter <= 1.6 and
  diameter >= 0.5 and
   material_type = high_speed_steel and
   (type_of_tool = brazed or
   type_of_tool = solid or
   type_of_tool = indexed))
```

This will return from the data base tools that match the specification. In case several tools are available, then heuristics are used to select the best one. Some of the heuristics follow:

- Choose the minimum cost tool.
- Choose the tool with with more tool life left.
- Choose the tool according to the location, as discussed earlier.
- Choose the tool with the maximum number in stock.

6.3.4 Operation Sequencing

One strong point of the AMPS planner is the ability to generate an optimal operation sequence. The word *optimal* is actually misleading. The sequence generated is near optimal for minimizing the number of setups, and minimizing the number of tool changes. Optimality is a difficult issue to address in the real manufacturing environment. We tried to find a good solution which satisfies all constraints. Constraints include static constraints and dynamic constraints. Static constraints are those constraints discussed in the feature refinement section. Dynamic constraints arise during the fixture planning; for example, a feature cannot be machined due to fixture interference, which produces a constraint on the feature. There are many feasible operation sequences which satisfy all constraints; only the best one is selected. The one selected uses the minimum number of setups and the minimum number of tool changes.

The sequencing is done by applying heuristics on a precedence diagram and a set of feature clusters. The precedence diagram represents all static constraints. As long as the precedence is followed, we are sure that all static constraints have been satisfied. Dynamic constraints are fed back from the fixturing method planner. They are considered only when one arises. A feature cluster is a group of features which have the same approach direction. A cluster therefore represents a candidate setup. An example of a precedence diagram and cluster set is shown in Fig. 6.18. More on precedence and clustering is discussed below.

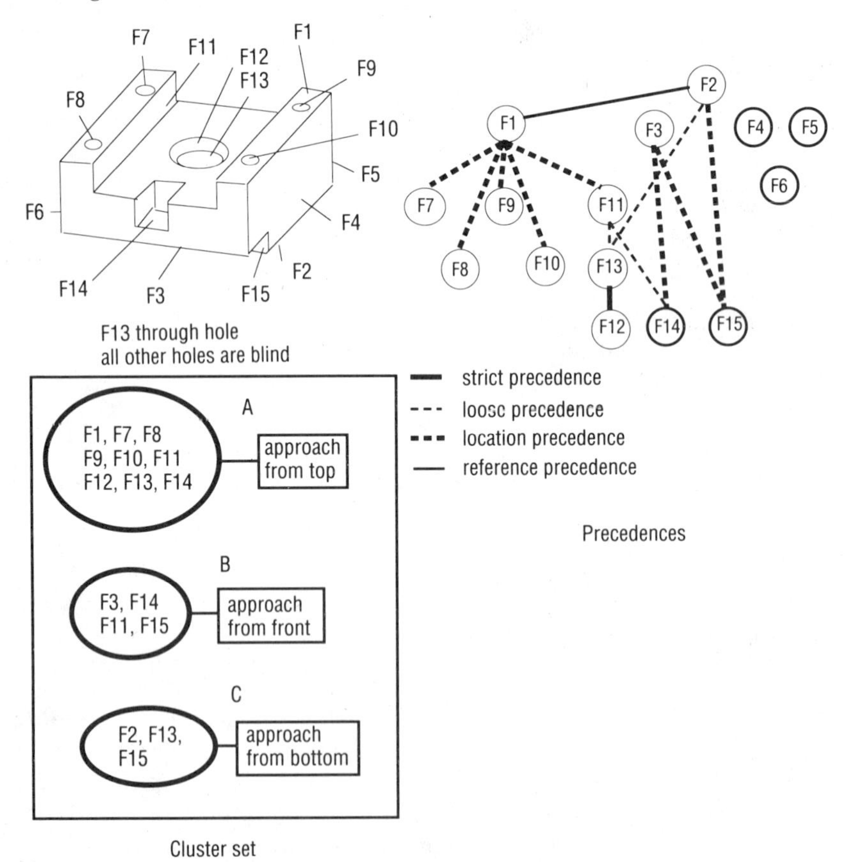

Figure 6.18 Sequence determination

6.3.4.1 Precedence creation

There are four types of precedences:

- Strict precedence due to process geometric constraint, such as boring must be preceded by a drilling processes
- Loose precedence due to good manufacturing practice, such as milling the slot before drilling the bottom hole
- Location constraints due to location tolerance
- Reference constraints due to the reference (datum) surface.

Among the precedences, strict and loose precedences are related to the geometry. Geometric data can provide the information needed for precedence determination. Rules have been written to generate these precedences. A sample rule is shown below:

```
((if (text (precedence progress))
  (?feat1 is in class hole)
  (the contains_b of ?feat1 is ?feat2)
  (?feat2 is in class hole)
  then
  (the loose_precedences of ?feat1 is ?feat2))))
```

The above rule establishes a loose precedence between two holes. Hole one (feature 1) contains hole two (feature 2) at its bottom.

A precedence due to reference might be created by a rule that searches for a face with the largest area. This selected face can be used as the reference for the subsequent manufacturing operations.

6.3.4.2 Clustering

Feature clustering is based on a commonality which can either be tool approach directions or tools. In case tool approach direction is used, all features which share the same approach direction can be grouped into one cluster. For a given part, there can be several clusters. Since a feature may have several approach directions, it may belong to several clusters as well. For example, in Fig. 6.18, there are three clusters A {F1, F7, F8, F9, F10, F11, F12, F13, F14}, B {F3, F11, F14, F15}, C {F2, F13, F15}. F11 and F14 belong to both clusters A and B. F15 belongs to both clusters B and C. The final setup is a refined cluster. Refining a cluster is to remove an element from a cluster or to regroup a cluster. When a feature requires more than one operation to finish, it may be represented as a different cluster element. Each cluster element appears in the final clusters once and only once.

6.3.4.3 Generating sequence

With precedence and clusters on hand, the operation sequence can be generated. There are two levels in the sequence determination: the global level and the setup level. The global level determines the final cluster and its sequence. The setup level sequencing determines the feasibility of the features to be machined in the particular setup. The operation sequence within the setup is also determined.

The global level sequencing is actually done before process selection and tool selection. The following operations are conducted:

Step 1. Create clusters using tool approach direction as the commonality.

Step 2. Refine clusters using the precedences.

Step 3. Apply heuristics to remove the duplicate features from clusters. A feature can appear in one cluster only. Current AMPS trys to minimize the number of clusters by maximizing the size of a cluster.

The setup level sequencing is done after tool selection.

Step 1. Create clusters using tool as the commonalty.

Step 2. Refine clusters using the precedences and operating sequence within the cluster determined.

Step 3. Check the precedence between the clusters, if there is any conflict, split the cluster and reorder the sequence between the clusters.

By creating clusters using tool as the commonality, the objective of minimizing tool change can be achieved. Steps 2 and 3 try to make the final sequence satisfy the precedence constraints. At this stage the final sequence is determined.

6.3.5 Fixturing Method Planning

The fixturing method planner is call CLAMPS [Shah 1988]. The purpose of CLAMPS is to generate a fixturing plan which ensures a firm holding of the workpiece and a minimum number of refixturings. Eventually, the fixturing method planner will use modular fixtures as the fixturing device. The current implementation considers only a vise with parallel jaws (Fig. 6.19). The fixturing method planner is incorporated into the process planning system. The data input includes features (refined features) in a setup, the design model, the intermediate workpiece boundary model, strict precedences generated by the feature refinement routines, and the "preferred" precedences generated by the process selection module.

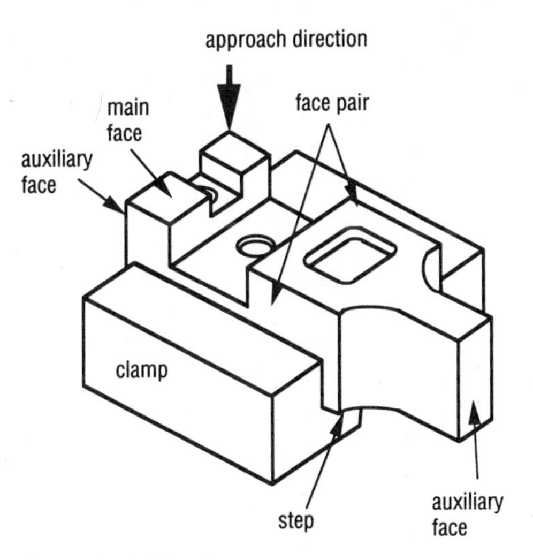

Figure 6.19 Fixturing

Output from the fixture planner is a list of features that can be machined taking the fixturing constraints, the orientation, and the position of the workpiece in the fixture into consideration.

The vise clamping module uses the following procedure to determine the fixturing method:

Step 1. Select from the workpiece boundary model all faces that are oriented perpendicular to the main approach direction V of the setup $\{F_i, i=1,n \mid N_i \times V = 0\}$, N_i is the normal of face F_i.

Step 2. Pair faces $FP_k = (F_i, F_j)$ with opposite surface normal ($N_i = - N_j$) and store them as candidate clamping faces.

Step 3. Delete from the candidate clamping face pair the ones that have face overlapping area A_k less than a certain acceptable value. Face overlapping area is defined as the intersection of a face projected to the other face along the normal direction.

Step 4. Calculate a composite weight W_k for each face pair FP_k. The composite include weights on overlapping area, inverse of the face distance, and number of features likely to be machined on the auxilliary faces (see Fig. 6.19).

Step 5. Select the face pair FP^* which has the maximum weight W_k. $FP^* \rightarrow W^* = \max \{W_k , k = 1,...m)$

Step 6. If there is a through hole present on or in the feature set of the setup, calculate the distance d from the feature to FP^* . This distance d is used as the maximum vise step size.

Step 7. Query the data with the face pair data and *d* for the appropriate vise jaw.

Step 8. Calculate offset of the workpiece for machining features on the auxiliary faces.

Cutting force and stability analysis can also be performed. The cutting force is estimated using fitted curves of empirical data. Depending on the tool and workpiece materials, tool geometry, and cutting parameters, the cutting force is predicted. This operation can verify the clamping. If it fails, then the clamping strategy needs to be modified. Finally, the potential interference between the fixture and the features to be machined is checked. If everything is satisfactory, the task is completed; otherwise, a new set of clamping faces is considered. An example of fixturing planning results is shown in Fig. 6.27 on page 226.

6.3.6 NC Cutting Path Generation

NC cutting path generation is the last step in process planning. At this stage, setups, fixturing method, processes, tools, and a preferred machining sequence have been determined. The next step is to finalize the machining sequence and to generate NC part programs automatically. It is necessary to generate a part program for each setup. Since there is no guarantee that the workpiece will be located in the machine work space exactly as planned, the probe on the machine tool is used to establish the reference. In order to generate the correct NC cutter path, it is sufficient just to consider individual features. The following problems need to be considered:

- For each cutting tool, re-consider the sequence of features to be machined.
- Generate movements to avoid collision with the workpiece and fixture.
- Determine the probing location in order to find the reference.
- Find the cutter path within a given feature.
- Output data in the required CL data format.

There are two types of motions in a cutting path: rapid traverse and feed motion. The rapid traverse is used to move the cutter from the end of one feature to the beginning of the second feature. This motion must not make the tool collide with anything. The other motion is the feed motion. The feed motion is the one does the actual cutting. Since the rapid traverse is usually at a rate of ~200 ipm, and the workpieces are smaller than a 10 inch cube, the travel time optimization is insignificant. The collision problem is the main issue that needs to be addressed. In case of feed motion, which usually has a rate of ~10 ipm, the path optimization is quite important.

The QTC system takes the following steps in CL data generation:

1. Evaluating the intermediate workpiece boundary model
2. Probe location determination
3. Movements and detours
4. Feature machining path generation
5. Code output

First the workpiece boundary model at each setup is evaluated. The intermediate workpiece model is needed for probing points determination. As mentioned before, the probe is used to establish the reference. Six points on three surfaces are probed. The probing surfaces are first determined on the raw workpiece model. The approach surface opposite to the machine spindle is one of the probing surfaces. The top surface and the side surface are two other candidates. On each surface two probing points are taken. The further apart these two probing points, the more accurate one can get. The original probing surface on the raw workpiece may be machined during a previous setup. The machining operation may split the original surface into several disjointed surfaces. There may be a set of surfaces to be considered for locating the two probing points. At each setup, the faces which have the tag of the original probing face are collected. The two extreme vertices from those faces are obtained through a search algorithm. The probing points are

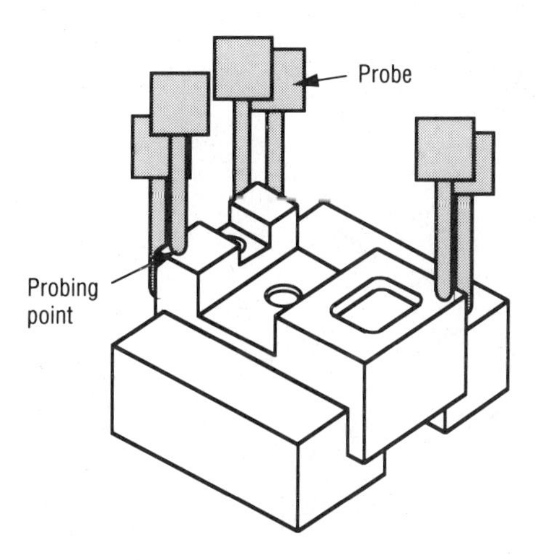

Figure 6.20 Probing of workpiece location

moving inward from the two extreme points in order to ensure there is no collision with the fixture and that the probe will touch the workpiece. The NC code for probing is output to the CL data file.

For the movements and detours generation, a safety envelope is constructed around the boundary model of the part. Any rapid traverse from one location on the part to another location is always via two points outside the safety envelope. This ensures that the rapid traverse is collision free. The via points are obtained by a positive z motion (move tool up) from the points on the part. Although this may not be the shortest path for rapid traverse, the method is simple and reliable. Also, as discussed earlier, even if one does find the optimal path, the time saving is negligible.

The actual machining of each feature is done by using a parametric cutter path model for each feature. For example, to drill a hole the parameters are depth of the hole, clearance plane height, hole location, and hole type (such as through hole, blind hole, etc.). The system also checks the depth/diameter ratio in order to set the strategy for chip removal. For example, when the depth/diameter ratio is greater than one, a multiple pass drilling path will be generated. The parameters are not directly taken from the design feature model; instead, the refined feature data is used. This is again a major feature of the QTC system. The cutter path models are written to be as generic (flexible) as possible. Therefore, they require much more input information than their counterparts used in turnkey CAD/CAM systems. Although not implemented in QTC yet, for example, a cutter path model developed for a pocket [Bala and Chang 1988] can cut a virtual pocket with multiple islands. A virtual pocket is a

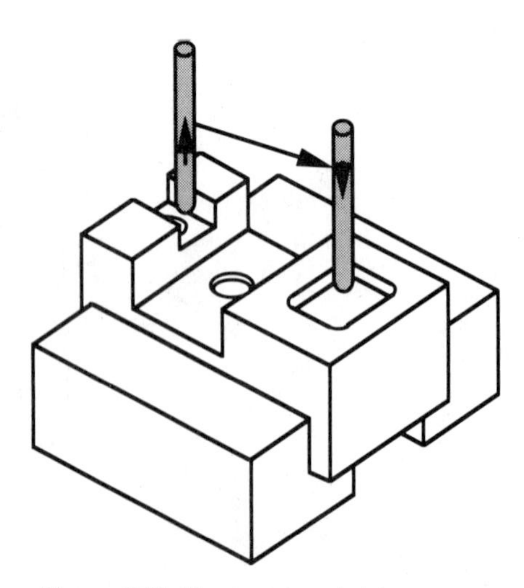

Figure 6.21 Movement and detour

generalized pocket which allows some of its boundary faces to be missing. For example, a through slot can be viewed as a virtual pocket with two missing end faces. The missing faces are called virtual faces which can be violated by the cutter. In contrast, the real faces cannot be violated. The feature refinement module supplies the exact dimension of the feature on the part.

In order to make intelligent cutter path generation, the geometry of the intermediate workpiece and the finished part are evaluated again. The intermediate workpiece geometry after each cut (defined as the removal of a feature) is important for determining how a tool can start cutting. For example, if slot 1 opens into slot 2, the approach direction for both is from the top (-Z direction [0 0 -1]); the through direction as well as feed direction for slot 1 is from both openings. The above refinement results are based on a finished part analysis. However, if slot 1 is machined first, the opening on slot 2 is not cut yet. Therefore, the only through direction is from the other opening. This can be detected through intermediate workpiece boundary evaluation.

Currently, the system does limited evaluation of workpiece geometry during the cutter path generation stage. It has the intelligence to minimize the cutter path when there are several intersecting slots. For hole making, it also checks the clearance beneath the hole. When due to the availability of the tool, a smaller cutter is selected, a multi-pass milling path will be generated for a slot.

6.3.7 Process Plan Documentation

The final result of the process planning function is a process plan documentation and several part programs written in a cutter location data format [Chang 1987]. These data files are used by the cell control module. In order to modularize the system, a process plan documentation language was developed. Using this language, all information pertinent to the machining of a part is available. This allows for independent future extensions to the process planning and cell control modules, without having to have format changes. Since the documentation contains detailed processing information, and it is machine understandable, it can be used in scheduling a flexible manufacturing cell as well.

A process plan documentation is divided into sections. Each section begins with a "%" character followed by a section code. Under each section there are subsections. The major sections, their codes, and subsections are the following:

%A Workpiece Information—identifier, technological information (material, hardness, etc.), CAD data pointer, inspection, cost data, and scheduling information

```
/******************************/
/*          CL data format definition          */
/******************************/
Comments 0 characters
Linear Intp 1 X          Y          Z
Circ Intp  2 plane  dir  X   Y  Z  X0  Y0  Z0
              (X0,Y0,&Z0 is the center)
              plane = 0 = XY plane
                      1 = XZ plane
                      2 = YZ plane
              dir = 0 CW
                    1 CCW
Feed        3  rate 0 = IPM/MMPM
                    1 = 1/T mode
Rapid       4 0- off
              1- on
Coolant     5 0- off
              1- flood (ext)
               2- mist (ext)
               3- flood (spindle)
               4- mist (spindle)
               5- pulse (spindle)
Tool        6 0- prepare   tool number
               1- load
Spindl       7  0 = off      rpm
               1 = clw
               2 = cclw
Cut Diam   8      cut diam
Cycle       9  0 = off
               1 = on
               2 = tap
               3 = mill
               4 = ream
               5 = bore  0 = normal
                         1 = dead spindle retract
                         2 = dwell
               6 - drill  0 = normal T = time of
                                         dwell
                         1 = dwell
Measurmnts 10  0 = inch
                1 = metric
```

/* note below, for all 2 digit mode codes, the second digit always refers to the axis being used, i.e. 1=X(and I),2=Y(and J), and 3=Z(and K). */

Figure 6.22 CL Data format

```
Probe 11 0 = disarm surface sense
         1 = arm surface sense
         2 = reset offset  H
         3 = find & measure
              /*change coordinate axis*/
                  0 1   X     I
                  0 2   Y     J
                  0 3   Z     K
                  0 4   Z     R
                  1 1   X
                  1 2   Y
                  1 3   Z
         4 = test presence
                  1   X
                  2   Y
                  3   Z

     /* offset H, as current tool tip position (G0)*/
     /* offset H, as the average of
                                measurements (G1)*/
     /* offset H, minimum measured coordinate
            the programmed axis value  (G2)  */
     /* offset H, has measured coordinate
            the programmed axis value  (G3)  */

         5 = compute offset
                0 = G0  0  H
                1 = G1  1  H   X
                2 = G2  2  H   Y
                3 = G3  3  H   Z
                4  H   X   Y
                5  H   X   Z
                6  H   Y   Z
                7  H   X   Y   Z
Index      12   deg rotate
Interrupt  13 0 = program halt  T = length of dwell
              1 = planned stop
              2 = dwell
Message 14   message string
Coordinates 15 0 = reset
               1 = absolute
               2 = increment
               3 = pre-load abs  X   Y  Z  B
Fixture Offset 16 0 = Off
                  1 = On       H
Program Labl 17  I8
```

Figure 6.22 CL Data format (continued)

```
/**************************************/
/*              Data field format            */
/**************************************/
Ixx    integer of xx length (max)
Fxx.x  fixed number of xx.x length (max)
Sxx    string of xx length (max)
[]     refers to an optional field which may or may not be
       present, based on the value of a previous field.
/**************************************/
                  FIELDS
Opp. group   2     3     4     5     6     7     8
/**************************************/
  0  S80
  1  F4.4 F4.4 F4.4
  2  I1 F4.4 F4.4 F4.4 F4.4 F4.4 F4.4
  3  F1.4 I1
  4  I1
  5  I1
  6  I1 I16
  7  I1 I4
  8  F1.3
  9  I1  [I1] [F2.2]
 10  I1
 11  I1  (0,1)
     I1   (2) I1
          (3) I1 I1 F4.4 [F4.4]
          (4) I1 F4.4
          (5) I1 I1  I1 [F4.4]   [F4.4]   [F4.4]
 12  I3
 13  I1 [F2.2]
 14  C60
 15  I1  [F4.4] [F4.4] [F4.4] [F4.4]
 16  I1
 17  I8
```

Figure 6.22 CL Data format (continued)

%B Shop Data—data base names, costs associated with the workpiece and machining

%C Process Plan Information—pre-loading information (m/c tool, heat treatment, mtl handling device and program, number of setups, etc.), machine tool information, cleaning and deburring, inspection

%D Machine Tool Level Description—setup information, clamping devices, locators, setup devices, setup time, workpiece orientation, setup CAD data, setup inspection, number of operations

%E Operation Level Information—machining specifications, operation tool number, workpiece orientation, operation data, operation inspection

Each subsection is also identified by two "%" and a two digit code. There are several fields within each subsection. A field is assigned a unique field name. Each section or subsection can appear multiple times. Since the documentation is keyword driven, it is easy to expand and to parse. The QTC system cell controller uses the documentation for scheduling tasks. An example process plan documentation can be found in Section 6.5.

Attached to each setup, under the machine tool level description, is an NC cutter location (CL) data file. The CL data file is generated by the NC cutter path generation module. An in-house developed CL data format is used initially. The format of the CL data file is given in Fig. 6.22. Post processors in the cell controller translate this data format into machine specific part program format. The post processor information is given in the machine tool level.

6.4 System Implementation

Both hardware and software environment were determined early in the project. Interfaces between software modules were also defined. The Sun 3 workstation was chosen as the hardware environment because it is a UNIX®-based machine and it supports Ethernet and TCP/IP protocol. It provides a superb development environment. UNIX is becoming the standard operating system for many machines ranging from mainframe computers to personal computers. By chosing an UNIX-based system, we can port our system to another hardware platform easier. Under the UNIX system, we adopted X-Window® (X version 10) [X 1986] as the windowing system. X was selected also because of its portability. Many vendors support X on their workstations. At the time of adoption, X-10 had just begun to be available. Although it allows us to open a window on a remote machine and provides many other convenient functions, the weak graphics (lack of three dimensional graphics, inadequate two dimensional graphics, lack of dialog box, and lack of a pull-down menu capability) forced us to write additional routines on top of the X package. Since summer 1988, X version 11 became available to us, and some of the advanced graphics functions have been implemented in X. However, the conversion from X-10 to X-11 was not as straightforward as we expected. Many function names have been changed from X-10 to X-11.

For the programming language, CommonLisp and C language were selected as the standard implementation languages. Again the rational is the portability. The KEE® [intellicorp] expert system shell was used initially for the process selection module. KEE® has provided us with a powerful knowledge modelling environment. The built-in inference engine also saved us time on system development. CommonLisp is naturally integrated into the KEE® system. For the many procedure codes which are written in C, a KEE®/C interface module from Intellicorp enables us to call C functions and return values back to KEE® . The returned values are placed into slots in a frame. The disadvantages of using an expert system shell as KEE® are the following:

- It requires a large amount of memory to run.
- It is slow to load and to run the code.
- It requires a different windowing system—Sun TOOLS®.

The original QTC system runs on a Sun 3/160 server with 20 M byte of RAM and two 380 M byte hard disk. The system load time is 15 minutes. In order not to disrupt the process, another Sun 3/50 is used to run the design and the cell control system under X window. The advantage of using an expert system shell for initial program development is the rich development environment offered by the shell. However, since the process planning system ideally should run as a background process, the majority of the human interface facilities (such as graphics) provided by the shell is not needed. The overhead on a shell is not justifiable by the convenience of using it. At the time of this writing, the expert system is being rewritten using CommonLisp. After the code is compiled, it should speed up the loading and execution of the system. The demand on memory will also be reduced.

As a research system, the current QTC implementation is adequate. In order to migrate the system onto the shop floor, the computer platform must be able to run alone. It has to be affordable to the small shop operator. The current arrangement with a Sun server and a slave cost more than $50,000 for industrial use. The KEE® package costs another $50,000. Fortunately, after KEE® is removed, the system should be able to run on a less than $10,000 machine. The second problem an industrial user has to face is the ease of use of the software. Better user interface design is warranted. However, these are things which can be done by any experienced computer programmer.

6.5 An Example Session

The following sequence of figures illustrate the operation of the QTC system. Figure 6.23 shows a design window. On the top of the window is a menu bar. From left to right there are file, edit, features, etc., menu items. A user uses a mouse to select command from these pull-down menus. First, pre-cut raw material of the desired size is selected from a database, then features are added. In the figure there are three design windows and a dialog box. The window on the lower right corner (the main window) displays the feature model with which the designer interacts. It can be seen that the feature model

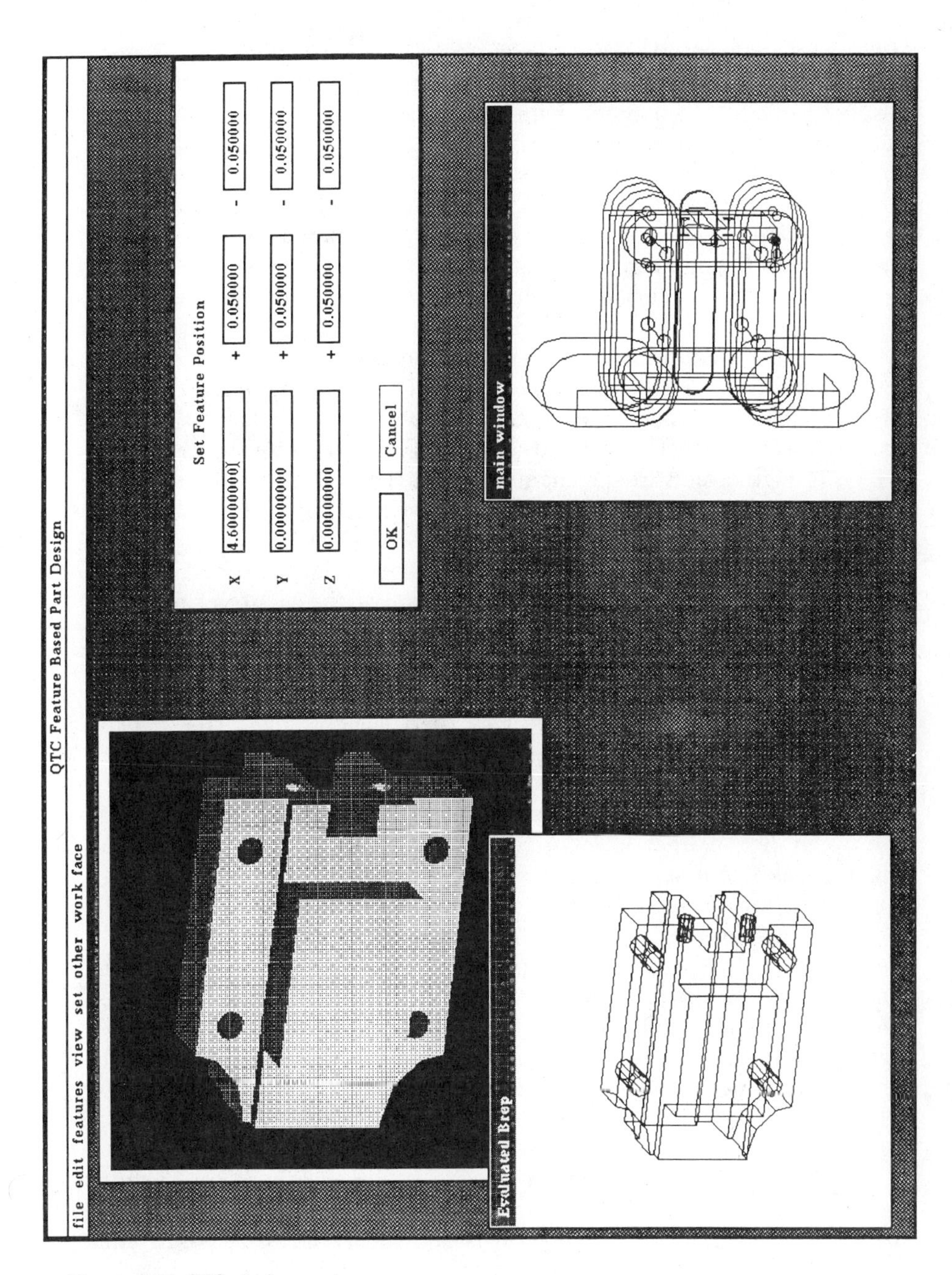

Figure 6.23 OTC design system

display is not very readable. After a few features have been added, there are lines and curves all over the display. A boundary model display can provide a much better picture. The lower left window shows an evaluated boundary model. At the time of this writing no hidden line removal algorithm has been implemented. The display is a simple wireframe. It is already much more readable than the feature model. In the upper left window is a shaded image of the design. This image is more realistic than the two others.

The design can be viewed from different directions by using the view menu. The windows can be sized and moved. Since engineering design requires accuracy, the design parameters, such as dimensions and tolerances, are entered through a dialog box (upper right corner of Fig. 6.23). The dialog box in the figure shows a dialog box for feature positioning. The current feature being positioned is the one which is highlighted in the main window. The black dot at the lower right corner of the part is one of the handle of the verticle slot. The displacement of the selected handle to the workface frame is entered as X, Y, and Z position. Also shown in the dialog box are default plus and minus tolerances. A user can change the value by moving the cursor to the selected tolerance box.

After the design is done, the process planning system is activated by entering the design data file name. As mentioned before, the process planning system uses KEE® expert system shell. The display in Fig. 6.24 shows the messages displayed during process planning. Although the process planning function can be done at the background, the display provides an interface for program debugging. On the upper right corner is an Active Area. It can be seen that setup planning and tool selection tasks are being conducted. For the selected face for set up—FACE3—several features are being machined. Tools are then selected for each feature. It shows that a query is generated by the process planning system and sent to the tool database. Below the active area is the area where messages are displayed. Since it does not scroll, the current task being done is the tool selection; the immediate prior task was sequencing. At the left of this area is the function area; it displays the functions being executed. The rest of Fig. 6.24 is not relevant and can be ignored. All the process planning tasks are done automatically.

After the complete process planning task is finished, the system generates a process plan documentation and a part program. A partial process plan documentation is shown in Fig. 6.25. A CL data file is generated for each setup. CL data for setup 3 is shown in Fig. 6.26. CL data is post-processed into a machine specific part program. The post-processor is activated by a cell controller.

The user interface for the cell controller is shown in Fig. 6.27. The screen is divided into four regions. At the top is a menu bar; it

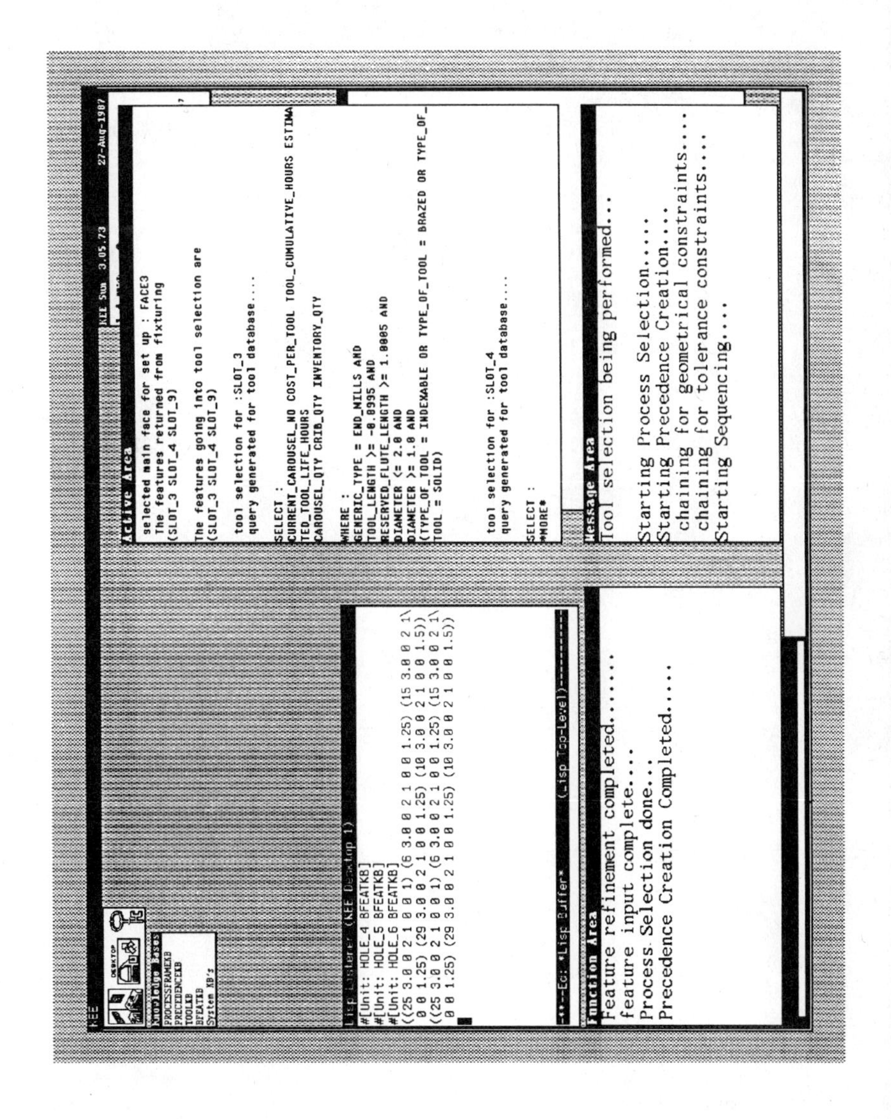

Figure 6.24 Process planning window

```
%A "WORKPIECE INFORMATION"

%%A1"PART"
PTNO 10 PTNA "DEMO"

%%A2"TECHNOLOGICAL"

MTDS "FMCS_WROUGHT_LC_RESULFURIZED"
WPTY P WBHN 0 WPSH P
WPLE 3.0 WPWI 5.0 WPHI 1.2

%C "PROCESS PLAN INFORMATION"

%%C1 "PRE-LOADING INFORMATION"
MTSQ 1
PMHD M  NOSP 3

%%C2 "MACHINE TOOL INFORMATION"

MCNO 1211212112 MCNA "T_10_CNC"

%D "MACHINE TOOL LEVEL DESCRIPTION"

%%D1 " SETUP INFORMATION"
STNO 1001
STFP "SETUP_0"

%%D6 "WORK PIECE ORIENTATION"
WPOP 15.684 7.460 8.296
WPOR 1.5708 0.000 3.141

%%D9 "NUMBER OF OPERATIONS"
NOOP 3

%E  "OPERATION LEVEL DESCRIPTION"

%%E1 "MACHINING SPECIFICATIONS"
OPNO 100101
OPDS PRB

%%E2 "OPERATION TOOL NUMBER"
OPTN 18

%%E4 "OPERATION DATA"
OPSR OFF OPCT OFF
OPFN "SETUP_0"

%E "OPERATION LEVEL DESCRIPTION"

%%E1 "MACHINING SPECIFICATIONS"
OPNO 100102
OPTY R OPDS MSL

%%E2 "OPERATION TOOL NUMBER"
OPTN 10

%%E4 "OPERATION DATA"
OPSP 104.7 OPDF 9.480469 OPDP 0.765
OPSR CLW OPCT FLE
OPFN "SETUP_0"

%E "OPERATION LEVEL DESCRIPTION"

%%E1 "MACHINING SPECIFICATIONS"
OPNO 100103
OPTY R OPDS MSL

%%E2 "OPERATION TOOL NUMBER"
OPTN 4

%%E4 "OPERATION DATA"
OPSP 104.1 OPDF 2.317128 OPDP 0.265
OPSR CLW OPCT FLE
OPFN "SETUP_0"
```

Figure 6.25 A partial process plan documentation

provides a pull-down menu for the user. On the right hand side are nine icons representing nine system functions. The bottom two windows (partially covered by two fixturing display windows) show system status. The rest of the screen is used to display graphics windows. In the figure, the design display icons are selected. Four fixturing display windows are selected under the design display functional module. The lower left window shows the raw workpiece. The notches on the lower vise jaw help the machine operator to position the workpiece. Above the raw workpiece display is a display of the workpiece after features in setup one have been machined. The lower right window shows a workpiece setup after setup two has been completed. Finally, the last window shows the finished part in the vise. The display of setups is user selectable. All pictures are generated by the system based on the process planning results.

```
15 0
11 2     1
1 25.0250  6.1221   12.0000
6 1      18
1 25.0250  6.1221   12.0000
11 1
4 0
3 150.000   1
1 25.0250  10.1064  12.0000
1 14.0376  10.1064  12.0000
1 14.0376  10.1064  11.9960
16 1     1
11 3     1     3      8.4960
11 5     1     3      1      9.4960
1 14.0376  10.1064  11.9960
11 1
1 14.0376  7.8136   11.9960
1 18.3304  7.8136   11.9960
16 1     1
11 3     1     3      8.4960
11 5     1     3      1      9.4960
1 18.3304  7.8136   11.9960
6 0      13
11 1
4 0
3 150.000   1
1 18.3304  10.1064  11.9960
1 12.5840  10.1064  11.9960
1 11.1840  10.1064  11.9960
1 11.1840  10.1064  8.6496
16 1     1
11 3     1     1      14.6840
11 5     1     1      1      13.6840
1 11.1840  10.1064  8.6496
11 1
1 11.1840  8.1536   8.6496
1 11.1840  8.1536   9.1424
16 1     1
11 3     1     1      14.6840
11 5     1     1      1      13.6840
1 11.1840  8.1536   9.1424
6 0      13
4 1
16 1     1
1 11.1840  8.1536   12.1424
15 3        11.1840 8.1536   12.1424  180
1 11.1840  8.1536   12.0000
6 1      13
1 25.0250  6.1221   12.0000
1 25.0250  8.9600   12.0000
1 17.1840  8.9600   12.0000
1 17.1840  8.9600   9.5960
7 1      641
3 9.539    1
5 1
9 6      0
1 17.1840  8.9600   -1.6175
1 0.0000   0.0000   9.5960
6 0      23
4 1
1 17.1840  8.9600   11.6960
1 17.1840  8.9600   12.0000
6 1      23
1 25.0250  6.1221   12.0000
1 25.0250  10.0600  12.0000
1 14.2840  10.0600  12.0000
1 14.2840  10.0600  9.5960
7 1      835
3 9.473    1
```

Figure 6.26 A partial CL Data file for setup 3

Finally the part programs are downloaded into the machine controller, the workpiece is loaded, and the machining process is started. The finished part is shown in Fig. 6.28.

6.6 Conclusions

In this chapter an integrated design/process planning system—QTC was introduced. The QTC system was designed with the goal of being used in a one-of-a-kind prismatic part machining environment. It has successfully shown that the concept discussed in this book is realizable. The system is able to generate a sound detailed process plan and NC cutter path automatically. What makes the system unique is its ability to generate a process sequence automatically and correctly. The system is by no means complete. At the time of this

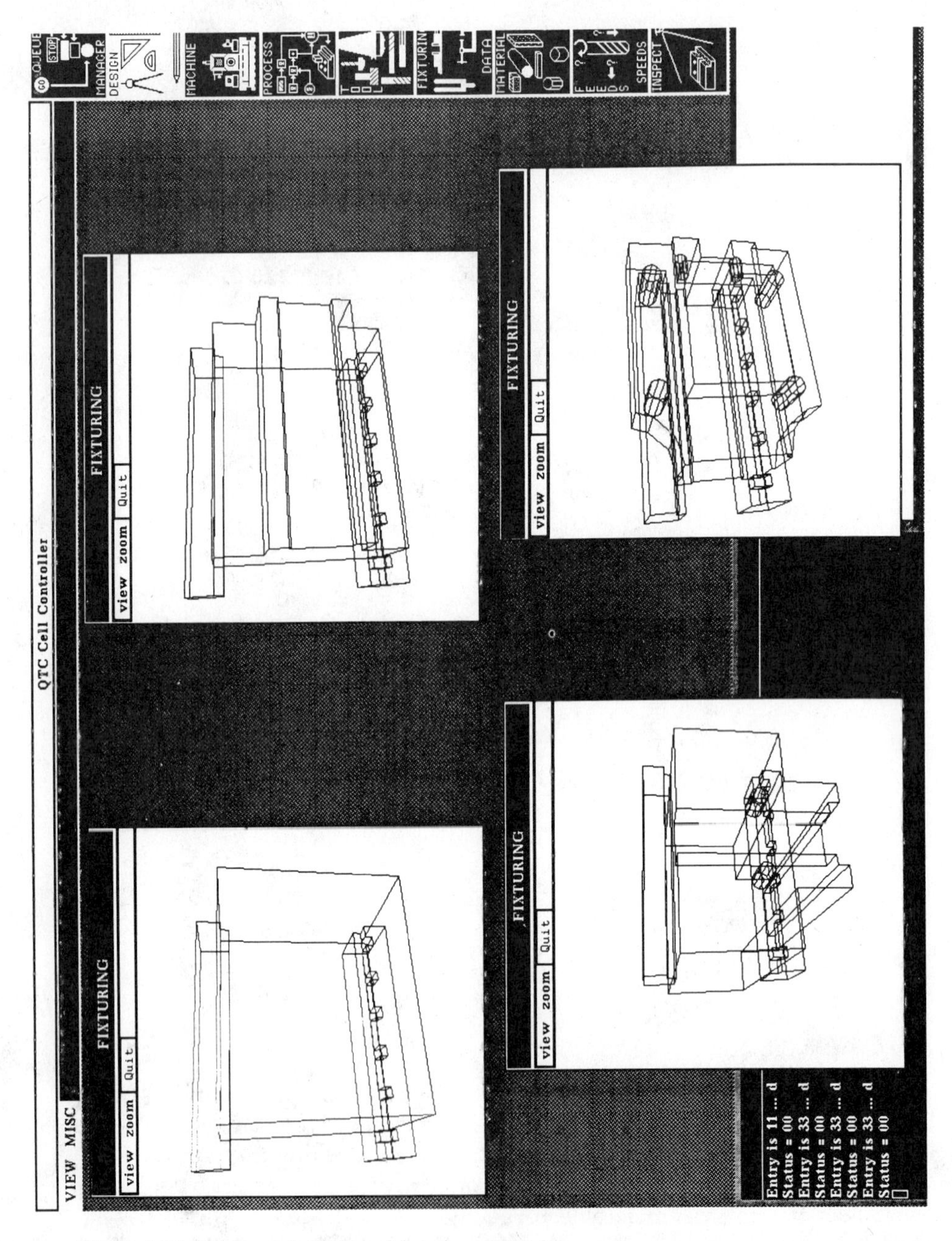

Figure 6.27 Cell control display of fixturing steps

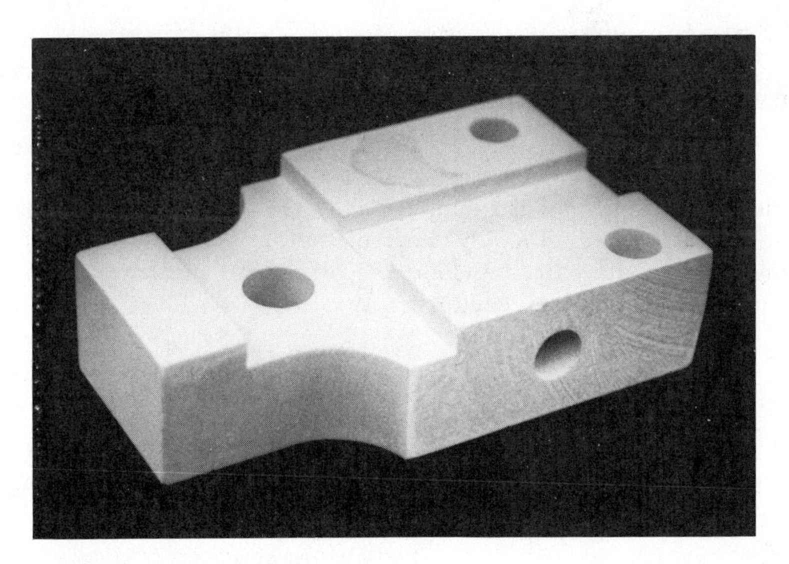

Figure 6.28 Finished part picture

writing many new features are being implemented. This general framework is able to accommodate the new additions.

However, the QTC system approach is not without some problems. The major problem is its complexity. To make a good process plan, one needs a lot of information and experience. When implemented in a computer system, this information and experience needs to be programmed in. From the discussion in this chapter, one may see that a tremendous amount of data and knowledge have been used in this prototype system. Data can be collected much easier than knowledge. To be useful, the knowledge has to be represented in the system. The major difficulty is the geometry related knowledge. Geometric reasoning is hard and tedious. There is no unified approach for all geometric reasoning problems, and we may never have one. Many of the problems have to be treated as special cases. Therefore, algorithms need to be developed for those special cases. It can be a very difficult task to develop all the necessary algorithms, especially when the domain of the features and processes attainable by the system increase. Some of the research issues which need immediate attention are listed below.

- efficient algorithm for feature refinement—Many of the algorithms implemented in the system are ad hoc in nature. More studies are needed in order to improve their efficiency.
- efficient and optimal algorithms for sequencing—Sequencing affects the quality and cost of machining. It is desirable to implement an efficient and optimal algorithm for the sequence determination.

- protrusion features—Protrusion features are design features. To manufacture a protrusion feature the material surrounding it is removed. To identify the removal volumes a full-fledged feature recognition scheme needs to be developed.
- curved surfaces—When curved surfaces are allowed to be part of the design, much of the simplicity the QTC system has enjoyed no longer exists. Although many of the methods developed for polyhedral features cannot be used, the major difficulty of having curved surfaces is with fixturing.

References

Chang, T.C. & Wysk, R.A., *An Introduction to Automated Process Planning Systems*, Prentice Hall, Englewood Cliffs, 1985.

Chang, T.C., "Integrated process planning approach for discrete part machining," Proceedings of 14th NSF Conference on Production Research and Technology, Ann Arbor, Mich., October 6–9, 1987, pp. 43–50.

Chang, T.C., "MAPT 3.0 user's manual," School of Industrial Engineering, Purdue University, W. Lafayette, Ind., 1987.

Choi, B.K., Barash, M.M., and Anderson, D.C., "Automatic recognition of machined surfaces from a 3D solid model," *CAD*, vol. 16, no. 2., 1984.

Date, C.J., *An Introduction to Database Systems*, Vols I & II, Addison Wesley, 1982.

Eversheim, W., Fuchs, H., and Zons, K.H., "Automatic process planning with regard to production by application of the system AUTAP for control problems," *Computer Graphics in Manufacturing Systems*, 12th CIRP International Seminar on Manufacturing Systems, Belgrade, 1980.

Henderson, M.R., "Feature recognition in geometry modeling," CAD-I's 13th Annual Meeting and Technical Conference, November 13–15, 1984.

Hummel, K.E. and Brooks, S.L., "Symbolic representation of manufacturing features for an automated process planning system," Proceedings, Symposium on Knowledge-Based Expert Systems for Manufacturing, ASME Winter Annual Meeting, Anaheim, Calif., December 7–12, 1986.

Joshi, S., and Chang, T.C., "Graph-based heuristics for recognition of machined features from a 3-D solid model," *Computer-Aided Design*, 1988.

Joshi, S., Vissa, N., and Chang, T.C., "Expert process planning system with solid model interface," *The International Journal of Production Research*, 1988.

Kanumury, M., "AMPS—An automatic manufacturing process planning system," M.S. thesis, School of Industrial Engineering, Purdue University, W. Lafayette, Ind., 1988.

Kanumury, M., and Chang, T.C., "Survey of process planning systems for turned parts," Engineering Research Center on Intelligent Manufacturing Systems, Purdue University, January, 1987.

Kramer, T.R. and Jun, J.-S., "Software for an automated machining workstation," Report, National Bureau of Standards, July, 1986.

Mashburn, T., "A polygonal solid modeling package," MS Thesis, Mechanical Engineering, Purdue University, 1987.

Mitchell, O.R., Lyvers, E.P., et al., "Recent results in precision measurements of edges, angles, areas, and perimeters," *Proc. of SPIE on Automated Inspection and Measurement*, vol. 730, Cambridge, Mass., October 1986, pp. 123–134.

Nau, D.S. and Chang, T.C., "Hierarchical representation of problem-solving knowledge in a frame-based process planning system," *Journal of Intelligent Systems*, vol. 1, no. 1, 1986, pp. 29–44.

Park, H.D. and Mitchell, O.R., "CAD based planning and execution of inspection," Proc. of the IEEE Computer Vision and Pattern Recognition Conference, Ann Arbor, Mich., June 5–9, 1988.

Shah, J., "CLAMPS—Automated fixturing in a flexible manufacturing environment," M.S. thesis, School of Industrial Engineering, Purdue University, W. Lafayette, Ind., 1988.

Shih, F.Y. and Mitchell, O.R., "Skeletonization and distance transformation by grayscale morphology," Proc. SPIE Symposium on Automated Inspection and High Speed Vision Architectures, Cambridge, Mass., 1987.

Wysk, R.A., "An automated process planning and selection program: APPAS," Ph.D. thesis, Purdue University, W. Lafayette, Ind., 1977.

Wysk, R.A., Chang, T.C., and Ham, I., "Automated process planning systems—an overview of ten years of activity," First CIRP Working Seminar on Computer Aided Process Planning, Paris, France, January 22–23, 1985.

X Window System Manual, MIT, in UNIX Programmer's Manual, 1986.

Bibliography

This bibliography is divided into seven sections:

1. general process planning,
2. variant process planning,
3. generative process planning,
4. expert system based process planning,
5. feature recognition in process planning,
6. automatic NC programming in process planning, and
7. fixturing method planning in process planning.

Since expert system-based process planning is also generative in nature, some articles are listed in both sections. For more articles published before 1982, refer to Chang and Wysk [1985].

A.1 General Background

Adlard, E.J., "The Use of Flexible Group Technology Codes in Process Planning," 19th NCS Annual Proceedings, 1982, pp. 154–161.

Alhgrim, S.C. and Chang, T.C., "A survey on the usage and development of computer aided process planning systems," Report to the Engineering Research Center on Intelligent Manufacturing Systems, Purdue University, February 1989.

Allen, D.K. and Smith, P.R., "Computer-aided process planning," Report of Computer-Aided Manufacturing Laboratory, Brigham Young University, October 15, 1980.

Allen, D.K., "An introduction to computer-aided process planning," *CIM Review*, Fall 1987.

Allen, D.K., "Process planning primer," Proceedings of the 19th CIRP International Seminar on Manufacturing Systems, Pennsylvania State University, June 1–2, 1987.

Allen, D.K., "Computer-aided process planning: software tools," PED-Vol 21, Integrated and Intelligent Manufacturing, (eds Liu, C.R. and Chang, T.C.), ASME, WAM, Anaheim, Calif., December 7–12, 1986, pp. 391–400.

Alting, L. and Zhang, H., "Computer aided process planning: the state-of-the-art survey," *International Journal of Production Research*, vol. 27, no. 4, 1989, pp. 553–585.

Alting, L., "A systematic approach in manufacturing," SME Technical Report, MS79-522, 1979.

American Machinist, "GT via automated CAPP," *American Machinist*, vol. 124 no. 4, 1980, pp. 119–120.

Bao, H., "CAPP and AUTAP: Two educational tools for teaching computerized process planning," *ATV-SEMAPP*, Technical Univeristy of Denmark, 1983.

Bond, V.A., "A conceptual model for integrating computer aided design and computer aided manufacturing," Master Thesis, The Pennsylvania State University, Pa, 1981.

Bucher, W.S., "An integrated group technology program," SME Technical Report, MS79-977, 1979.

Burge, W., "Integrating design, process planning and numerical control through the use of standard programs and the computer (CIM)," Proceedings of Autofact 4 Conference, November 30–December 2, 1982.

Butterfield, W.R., Green, M.K., Scott, D.C. and Stoker, W.J., "Part features for process planning," Report R-86-PPP-01, CAM-I, Inc., November 1986.

Chang, T.C. and Joshi, S., "Computer-aided process planning: an introduction," Automated Factory Handbook: Technology and Management, (eds D. I. Cleland and B. Bidanda), TAB Books Inc., 1989.

Chang, T.C. and Wysk, R.A., *An Introduction to Automated Process Planning Systems*, Prentice-Hall, Englewood Cliff, 1985.

Chang, T.C., "Computer-aided process planning today and in the future," Procedings IIE Fall Conference, November 13–16, 1983, pp. 349–355.

Chang, T.C., "Just in time information generation—a study of automatic process planning," 15th NSF Grantees Conference on Production, research, and Technology, January 8–13, 1989.

Chang, T.C., Lecture note on Computer-Aided Process Planning, Workshop on Information and Productivity, National Chiao Tung University, July 1986. (210 pages)

Chang, T.C., "The advances of computer-aided process planning, technical report," NBS-GCR-83-441, The National Bureau of Standards, U.S. Department of Commerce, July 1983., (126 pages)

Chen, M.F. and Ito, Y., "Investigation on the engineer's thinking flow in the process planning of machine tool manufacturer", Proceedings of 13th NAMRC, 1985.

Cheung, Y.P. and Dowd, A.L., "Artificial intelligence in process planning," *Computer Aided Engineering*, vol. 5, no. 4, August 1988, pp. 153–156.

Church, Janis, "Computer aids for assembly planning present and future," Proceedings of Autofact West, vol. 2, Society of Manufacturing Engineers, November 17–20, 1980, pp. 109–122.

Curtis, M.A., *Process Planning*, John Wiley, New York, 1988.

Davies, B.J., "Expert systems in manufacturing," Intelligent Manufacturing System, II., 2nd International Summer Seminar, (Milacic, V.R. ed.), August 24–29, 1987, pp. 87–92.

Doran, J.M. and W.S. Nelson, "Computer aided process planning and work measurement," 15th NCS Annual Proceedings, 1978.

El-Midany, T.T. and El-Tamimi A.M., "Computer aided process planning systems and approaches," IFAC Conference, 1982, pp. 537–554.

Evans, S., "Financial justification of state-of-the-art investment: a study using CAPP," Engineering Management International, 1982.

Eversheim, W. and Schulz, J., "Survey of computer aided process planning systems," *Annals of the CIRP*, 1985, pp. 607–611.

Eversheim, W., Koenig, W., Wesch, H. and Dammer, L., "Computer-aided planning and optimisation of cutting data, time and cost," *CIRP Annals*, vol. 30 no. 1, 1981.

Gott, B., "Catering for CAPP," *Computer Manufacturing*, April 1988, pp. 20–22.

Gupta, T. and Ghosh, B.K., "A survey of expert systems in manufacturing and process planning," *Computers in Industry*, vol. 11, no. 2, January 1989, pp. 195–204.

Haas, M. and Chang, T.-C., "A survey on the usage of computer aided process planning systems in industry," Engineering Research on Intelligent Manufacturing Systems, Purdue University, January 1987.

Halevi G. and Weill, R., "Development of flexible optimum process planning procedures," *CIRP Annals*, vol. 29, no. 1, 1980.

Halevi, G., "Production planning /module for all embracing technology," Proceeding of Autofact West, CAD/CAM V. III Anaheim, Calif., November 1980.

Ham I. and Bond, V.A., "Computer aided process planning," Working Paper, the Pennsylvania State University, 1980.

Ham, I. and McClenahan, D.G., "Computer optimization of machining conditions for shop production using the performance index method," SME Technical Paper MS73-174, 1973.

Ham, I. and Lu, S. C.-Y., "Computer-aided process planning: the present and the future," *Annals of CIRP*, vol. 37, no. 2, 1988, pp. 591–601.

Ham, I. and Shunk, D., "GT survey results/related to I- Program," SME Technical Report, MS79-348, 1979.

Hancock, T.M., "Effects of adaptive process planning on job cost and lateness measures," *International Journal of Operations Production Management*, vol. 8, no. 4, 1988, pp. 34–49.

Harvey, R.E., "CAPP: critical to CAD/CAM success," *Iron Age*, September 1983.

Hatvany, J., "What is economic, and how am I to know?" Proceedings of the 19th CIRP International Seminar on Manufacturing Systems, Pennsylvania State University, June 1–2, 1987, pp. 5–8.

Hitomi, K., *Manufacturing Systems Engineering*, Taylor & Francis Ltd., London, 1979.

Holtz, R.D., "GT and CAPP cut work-in-process time by 80%," *Assembly Engineering*, vol. 21, July 1978, pp. 16–19.

Houtzeel, A., "The many faces of group technology," *American Mechanist*, January 1979, pp. 115–120.

Jackson, R.H., "Automated process planning: the key to automating manufacturing support," The CASA/SME Tulsa Conference, March 1978.

Jackson, R.H., "Automated process planning: key to automating manufacturing support," SME Technical Report MS80-7834M, 1980.

Jiang, W., and Xu, H., "Capp systems and applications in China," Proceedings of the 19th CIRP International Seminar on Manufacturing Systems, Pennsylvania State University, June 1–2, 1987.

Johnson, B.T., "Expert systems in process planning," Proceedings of 4th European Conference Automated Manufacturing, May 1987, pp. 497–506.

Kanumury, M. and Chang, T.C., "Survey of process planning systems for turned parts," Engineering Research Center on Intelligent Manufacturing Systems, Purdue University, January, 1987.

Knutsen, B.H., "The need for process planning in mechanical assembly in general," *CIRP Annals*, vol. 22 no. 1, 1973.

Koenig, D.T., "Computer aided process planning for computer integrated manufacturing," Proceedings of the 19th CIRP International Seminar on Manufacturing Systems, Pennsylvania State University, June 1–2, 1987, pp. 127–130.

Kumara, S., Joshi, S., Moodie, C.L., Kashyap, R.L., and Chang, T.C., "Expert systems in industrial engineering," *International Journal of Production Research*, vol. 24, no. 5, 1986, pp. 1107–1126.

Kyttner, R., Shtcheglov, N. and Kimmel, A., "A complex computer-aided process—planning and optimisation for machine production," SME Technical Report, MS79–158, 1979.

Logan, F. A., "Process planning-the vital link between design and production," *Computer Aided Process Planning*, (ed J. Tulkoff) 1985.

Lyons, J. W., "The role of process planning in computer integrated manufacturing," 7th International Conference on the Computer as a Design Tool, London, UK, September 2–5, 1986.

Marsh, J., "Bridge between design and production (Computer-Aided Process Planning)," *Engineering Computing*, vol. 7, no. 6, September 1988, pp. 43–44.

Marshall, P., "Computer-aided process planning and estimating as part of an integrated CAD/CAM system," *Computer-Aided Engineering Journal*, October 1985.

Marshall, P., "The requirements of an integrated CAD/CAM system," *Computer-Aided Engineering Journal*, February 1985.

McDonald, J.A., "CAPP to support group technology," 28th Annual Proceedings, Standard Engineers Society, INC., 1979.

Merchant, E.M., "CAPP in CIM-integration and future trends," Proceedings of the 19th CIRP International Seminar on Manufacturing Systems, Pennsylvania State University, June 1–2, 1987.

Merchant, E.M., "The computer-integrated manufacturing as the basis for the factory of the future," *Robotics and Computer-Integrated Manufacturing*, vol. 2, no. 2, 1985, pp. 89–99.

Mill, F.G. and Spraggett, S., "Artificial intelligence for production planning,"*Computer-Aided Engineering Journal*, December 1984.

Mills, M., "A totally integrated system approach to design and manufacturing at McDonnel Douglas Corporation," Design Automation Conference, 1981.

Moseng, B., "The process planner's work place today and tomorrow," *Robotics and Computer-Integrated Manufacturing*, vol. 1, no. 3/4, 1984, pp. 237–244.

Niebel, B.W., "Mechanized process selection for planning new designs," ASTME Paper 737, 1965.

Nilson, E.N., "Integrating CAD and CAM-future directions," The CASA/SME Advanced Computer Techniques for Design and Manufacturing Conference, April 1977.

Nolen, J., *Computer-Automated Process Planning for World-Class Manufacturing*, Marcel Dekker, New York, 1989.

OIR Europe, Inc. "Group technology as CAD/CAM integrator in the 80's," Brochures and Information material, OIR Europe, Inc., 1986.

Pekelman, D., "On optimal utilisation of production processes," Operations Research, vol. 27, no. 2, March–April 1979, pp. 260–278.

Pinte, J., "Computer aided process planning," CAD/CAM 87, Teknologisk Institut, Denmark, October 20–22, 1987.

"Process planning software enhances accuracy and consistency," Special Report 795, American Machinist and Automated Manufacturing, June 1987.

Rahman, M. and Narayanan, V., "An expert system for process planning," \m Robotics & Computer-Integrated Manufacturing, vol. 3, 1987, pp. 365–372.

Reodecha, M. and Bao, H., "Development of a Classification and Coding System for Electronic Components: The First Step Toward Automatic Process Planning for Printed Wiring Assemblies," PED-Vol 19, Computer-Aided/Intelligent Process Planning (eds Liu, C.R., Chang, T.C. and Komanduri, R.), ASME WAM, Miami Beach, Fla., November 17–22, 1985, pp. 167–176.

Singal, K. and Weeks, J.K., "A compatibility matrix algorithm for generating alternate process designs," Working paper KS/JW I-A, Nov. 1979, University of North Carolina-Greensboro.

Sohlenius, G. and Kjellberg, T., "Artificial intelligence and its potential use in the manufacturing system," *Annals of the CIRP*, vol. 35, no. 2, 1986.

Srihari, K. and Greene, T.J., "Computer-Aided Process Planning—a CAD/CAM Interface," *CIM Review*, vol. 4, no. 4, Summer 1988, pp. 47–54.

Srinivasan, R. and Liu, C.R., "Evolutionary trends in generative process planning," Lecture Notes in Computer Science, CREST Advanced Course in Computer Integrated Manufacturing, Karlsruhe, FRG, September 1983.

Steudel, H.J. and Tollers, G.V., "A decision table based guide for evaluating computer-aided process planning systems," PED-Vol 19, Computer-Aided/Intelligent Process Planning (eds Liu, C.R., Chang, T.C. and Komanduri, R.), ASME WAM, Miami Beach, Fla., November 17–22, 1985, pp. 109–120.

Steudel, H.J., "Computer-aided process planning: past, present, and future," *International Journal of Production Research*, 1984, pp. 253–265.

Subramanyam, S., Lu, S.C-Y. and Zdeblick, W.J., "A characterization of the process planning task from an artificial intelligence perspective," Proceedings of the 19th CIRP International Seminar on Manufacturing Systems, Pennsylvania State University, June 1–2, 1987, pp. 197–206.

Tipnis, V.A., Vogel, S.A. and Gegel, C.L., "Economical models for process planning," 6th NAMRC, 1978.

Tipnis, V.A., "Computer aided process planning for aircraft engine rotational parts," SME Technical Report, MS79-155, 1979.

Traughber, T.J., "Software design considerations for computer aided process planning (CAPP)," ISATA 86, 15th International Symposium on Automotive Technology and Automation, Flims, Switzerland, 6–10 October. 1986.

Trusky, H.T., "Automated planning reduces costs," *American Machinist*, April 1984.

Tulkoff, J., (ed.) "Computer aided process planning" *SME*, 1985.

Tulkoff, J., "Process planning in the computer age," Machine and Tool Blue Book, November 1981.

Tulkoff, J., "Process planning in the computer-integrated factory," *CIM Review*, Fall 1987.

Tulkoff, J., "Process planning: An historical review and future prospects," Proceedings of the 19th CIRP International Seminar on Manufacturing Systems, Pennsylvania State University, June 1–2, 1987.

van Houten, F.J.A.M., "The development of a technological processor as a part of a workpiece programming system," *Annals of the CIRP*, vol. 30, no.1, 1981.

Vogel, S.A., "Use of group technology in process planning," Computer Applications in Manufacturing Systems, Devries, W.R. (editor), the American Society of Mechanical Engineering, 1980.

Waldman, H., "Process planning at sikorsky," CAD/CAM Technology, Summer 1983.

Wang, H.P. and Wysk, R.A., "Computer aided process planning taxonomies," 1988 Integrated Systems Conference Proceedings, Institute of Industrial Engineers, 1988, pp. 49–53.

Waterbury, R., "Computer assisted process planning—key to cost saving," *Assembly Engineering*, June 1980.

Wechster, D.B., "Process planning in Transition," *CIM Technology*, Winter 1984, pp. 31–32.

Weck, M. and Kiratli, G., "Applicability of expert systems to flexible manufacturing," *Robotics and Computer-Integrated Manufacturing*, vol. 3, no. 1, 1987, pp. 97–103.

Weill, R., Spur, G., and Eversheim, W., "Survey of comuter-aided processing planning systems," *CIRP Annals*, vol. 31 no. 2, 1982, pp. 539–551.

Willis, R.G. and Garrahy, J.R., "Measuring the performance of a manufacturing planning and control system," AUTOFACT West, CAD/CAM V. III, Anaheim, Calif., November 1980.

Wolfe, P.M., "Computer-aided process planning is link between CAD and CAM," *Industrial Engineering*, vol. 17, no. 8, 1985.

Wysk, R.A., Chang, T.C. and Ham, I., "Automated process planning systems—an overview of ten years of activity," Proceedings, First CIRP Working Seminar on Computer Aided Process Planning, January 22–23, 1985.

Zandin, K.B., "Computer aided process planning and productivity," The CASA/SME 1982 International Tool and Manufacturing Engineering Conference, May 1982.

A.2 Variant Systems

Abadi, A.A., "Computer-aided process selection and planning," unpublished Master Thesis, the Pennsylvania State University, 1984.

CAM-I, Inc., "CAPP 2.1 User's Manual," #PS-76-ppp-03, CAM-I Inc., TX., 1976.

Claytor, R.N., "CAM-I's CAPP System," SME Technical Report, MS76-241, 1976.

El Gomayel, J. and Nader, V., "Optimization of machine setup and tooling using the principles of group technology," *Computer and Industrial Engineering*, vol. 7, no. 3, 1983, pp. 187–198.

Emerson, C. and Ham, I., "An automated coding and process planning system using a DEC PDP-10," *Computers and Industrial Engineering*, vol. 6, no. 2, 1982.

Emerson, C.D., "Computer-coding and process planning," Master Thesis, the Pennsylvania State University, 1980.

Granville, C.S., "The role of GT/CAPP in a CIM Solution," Electro/88 Conference Record, May 10–12, 1988.

Groppetti, R. and Semeraro, Q., "CAPP-computer aided process planning using relational databases," *ISATA*, October 6–10, 1986.

Hayden, R.D., "Computer Assisted Process Planning and MIPLAN System," O.I.R. Inc., White Plains, N.Y.

Horvath, M., "Semi-generative process planning for part manufacturing," SME Technical Report, MS79-153, 1979.

Houtzeel, A., "Computer assisted process planning a first step towards integration," Proceedings of Autofact West, Society of Manufacturing Engineers, November 17–20, 1980.

IBM INC., "Automated manufacturing planning," Technical Report GE 20-0146-0, IBM Inc., White Plains, N.Y.

Ishizu, S., Furukawa, O. and Morimasa, S., "A methodology for an adaptive process planning," Department of Industrial and Systems Engineering, University.

Keytack, H.O. and Gravel, W., "Group technology applications in manufacturing operations through the computer system," Proceeding of the 19th CIRP International Seminar on Manufacturing Systems, Pennsylvania State University, June 1–2, 1987.

Lesko, J.F., "MIPLAN implementation at union switch and signal," The Association for Integrated Manufacturing Technology, 20th Annual Meeting and Technical Conference, April 1983.

Link, C.H., "CAPP, CAM-I automated process planning system," Proceedings of the 1976 NC Conference, CAM-I. Inc., Arlington, Tex., 1976.

Link, C.H., "CAPP-CAM-I automated process planning system", Proceedings of the 13th Numerical Control Society Annual Meeting and Technical Conference, March 1976.

Link, C.H., "Computer-aided process planning (CAPP)," SME Technical Report, MS78-213, 1978.

Logar, B. ad Peklenik, J., "Computer-aided selection of reference parts for GT-part families," Proceedings of the 19th CIRP International Seminar on Manufacturing Systems, Pennsylvania State University, June 1–2, 1987, pp. 131–135.

Milacic, V.R., et al., "SAPT-expert system based on hybrid concept of group technology," Proceedings of the 19th CIRP International Seminar on Manufacturing Systems, Pennsylvania State University, June 1–2, 1987.

O.I.R. Inc, "Introduction to MIPLAN," O.I.R., Inc., Waltham Mass.

OIR, MULTIPLAN, Organization for Industrial Research, Inc., Waltham MA, 1983.

Phillips R.H. and ElGomayel, Y.I., "Computerized process planning for metal cutting," 8th NAMRC, 1980.

Phillips, R.H., "A computerized process planning system based on component classification and coding," Ph.D. Dissertation, Purdue University, Indiana, 1978.

Rasch, F.O., "IPROS-a variant process planning system," Proceedings of the 19th CIRP International Seminar on Manufacturing Systems, Pennsylvania State University, June 1–2, 1987, pp. 157–160.

Schaffer, G.H., "GT via automated process planning," *American Machinist*, May 1980.

Scheck, D.E., "Feasibility of automated process planning," Ph.D. Thesis, Purdue University, West Lafayette, Ind., 1966.

Schwartz, S.J. and Shreve, M.T., "GT and CAPP productivity tools for hybrids," The *International Journal for Hybrid Microelectronics,* vol. 5, no. 2, 1982.

Simon, R. L., "Converging technologies shape the next generation of group technology and computer-aided process planning systems," PED-Vol. 21, The Winter Annual Meeting of the ASME, 1986, pp. 375–389.

Spur, G., Anger, H.M., Kunzedorf, W., and Stuckmann, G., "CAPSY—A dialogue system for computer aided manufacturing process planning," International Machine Tool Design Research Conference, vol. 21, 1978, pp. 281–288.

Tulkoff, J., "CAM-I automated process planning (CAPP) system," 15th NCS Annual Proceedings, 1980, pp. 304–324.

Waldman, H., "Process planning at Sikorsky," Computer Aided Process Planning, (ed. J. Tulkoff), *SME,* 1983.

Wang, H.-P. and Wysk, R.A., "Applications of microcomputers in automated process planning," *Journal of Manufacturing Systems,* vol. 5, no. 4, 1986.

Wang, H.P., "Microcomputer based process planning systems," Master's Thesis, the Pennsylvania State University, 1985.

Zandin, K.B., "Computer aided process planning and productivity," SME's 1982 International Tool & Manufacturing Engineering Conference, ion, Standards May 17–20, 1982, MS82-241.

Zdeblick, W.J., "CUTPLAN/CUTTECH a hybrid computer aided process planning and operation planning system," Proceedings of the 1st CIRP Seminar on CAPP, Paris, France, January 22–2, 1985.

Zhang and Gao, W.D., "TOJICAP—A system for computer aided process planning for rotational parts," *Annals of CIRP,* vol. 34, no. 1, 1984, pp. 299–301.

A.3 Generative Systems

Ajmal, A., "The development of a computer-aided process planning and estimating system for use in a jobbing foundry," Proceedings, the IXth International Conference on Production Research, August 17–20, 1988, pp. 589–597.

Ali, D.L., Ali, K.S. and Ali, A.L., "Neural networks: a feasible approach for process planning automation," Proceedings of the European Simulation Multiconference, June 1–3, 1988, 107–111.

Allen, K., "Generative process planning system using DCLASS information system," Monograph 4, Computer Aided Manufacturing Laboratory, Brigham Young University, 1979.

Alting, L., et al., "XPLAN-an expert process planning system and its further development," 27th International MATADOR Conference, April 20–21, 1988.

Anderson, D.C. and Chang, T.C., "Geometric reasoning in feature based design and process planning," *International Journal of Computers and Graphics,* (in press).

Austin, B.L., "Computer aided process planning for machined cylindrical metal parts," Proceedings of Autofact '86 Conference, November 12–14, 1986.

Austin, B.L., "Computer aided process planning for machined cylindrical parts," Bound Volume, Integrated and Intelligent Manufacturing, ASME Winter Annual Meeting, 1986, pp. 359–366.

Barkocy, B.E. and Zdeblick, W.J., "A knowledge-based system for machining operation planning," Proceedings of Autofact 6 Conference, October 1–4, 1984.

Barkocy, B.E. and Zdeblick, W.J., "A knowledge based system for machining operations planning," SME Technical Paper MS84-716, 1984.

Berenji, H.R. and Khoshnevis, B., "Use of artificial intelligence in automated process planning," *Computers in Mechanical Engineering,* September, 1986, pp. 47–55.

Berra, P.B. and Barash, M.M., "Investigation of automated process planning and optimization of metal working processes," Report 14, Purdue Laboratory for Applied Industrial Control, July 1968.

Boerma, J.R., "Fixes, a system for automatic selection of set-ups and design of fixtures," University of Twente, Laboratory of Production Engineering, January 1988.

Bowden, R. and Browne, J., "ROBEX—An artificial intelligence-based process planning system for robotic assembly," Proceedings, the IXth International Conference on Production Research, August 17–20, 1987, pp. 868–874.

Brooks, S.L., "Applying artificial intelligence techniques to generative process planning systems," M.S. thesis, University of Kansas, May 1986.

Brown, P.F. and McLean, C.R., "Interactive process planning in the AMRF," ASME Winter Annual Meeting, ASME, December 1986, pp. 245–262.

Brown, P.F. and Ray, A., "Research issues in process planning at the national bureau of standards," Proceedings of the 19th CIRP International Seminar on Manufacturing Systems, Pennsylvania State University, June 1–2, 1987, pp. 111–119.

CAD Centre Limited, "C-PLAN brochures and information material," CAD Centre Ltd., 1987.

CAM-I, "Conceptual model of automated process planning for the machined parts domain," CAM-I Process Planning Program, December 1988, pp. 1–10.

Chang, T.C. and Wysk, R.A., "Integratng CAD and CAM through automated process planning," *International Journal of Production Research*, vol. 22, no. 5, 1984, pp. 877–894.

Chang, T.C. and Wysk, R.A., "An integrated CAD/automated process planning system," *AIIE Transactions*, vol. 13, no. 3, September 1981, pp. 223–233.

Chang, T.C. and Wysk, R.A., "Automated process planning for automated manufacturing systems," First International and Fifth National Conference of Computers and Industrial Engineering, March 10–12, 1982.

Chang, T.C. and Wysk, R.A., "CAD/generative process planning with TIPPS," *Journal of Manufacturing Systems*, vol. 2, no. 2, 1983, pp. 127–135.

Chang, T.C. and Wysk, R.A., "CAD/generative process planning with TIPPS," Computer-Aided Process Planning, (J. Tulkoff, ed.), Society of Manufacturing Engineers, 1985.

Chang, T.C., Anderson, D.C. and Mitchell, O.R., "QTC—an integrated design/manufacturing/vision inspection system for prismatic part," Proceedings of the ASME 1988 Computers in Engineering Conference, Volume 1, July 31–Aug. 3, 1988, pp. 417–426.

Chang, T.C., "Integrated process planning approach for discrete part machining," proceedings of 14th NSF Conference on Production Research and Technology, October 6–9, 1987, pp. 43–50.

Chang, T.C., "Interfacing CAD and CAM—a study in hole design," Masters Thesis, Virginia Polytechnic Institute and State University, 1980.

Chang, T.C., "Quick turnaround cell—an integrated manufacturing cell with process planning capability," Control, Programming and

Integration of Manufacturing, (eds. D. Williams and P. Rogers), Butterworths, 1990.

Chang, T.C., "TIPPS—a totally integrated process planning system," Ph.D. Thesis, Virginia Polytechnic Institute and State University, 1982.

Chang, T.C., Wysk, R.A. and Davis, R.P., "Interfacing CAD and CAM—a Study in hole design," *Computers and Industrial Engineering*, vol. 6, no. 2, 1982, pp. 91–102.

Choi, B.K., "CAD/CAM compatible tool oriented process planning system," Ph.D. Thesis, Purdue University, 1982.

Chryssolouris, G., and Chan, S., "An integrated approach to process planning and scheduling," *Annals of CIRP*, vol. 134, no. 1, 1985.

Chryssolouris, G., and Gruenig, I., "Process planning interface for intelligent manufacturing systems," Proceedings of the 19th CIRP International Seminar on Manufacturing Systems, Pennsylvania State University, June 1–2, 1987.

Chryssolouris, G., et al., "Decision making on the factory floor: An integrated approach to process planning and scheduling," *Robotics and Computer-Integrated Manufacturing*, vol. 1, no. 3/4, 1984, pp. 315–319.

Crookall, J.R. and Smith, J.A., "Program autoplan for the derivation of operations and machining conditions in multi-operation turning," Proceedings of the Second International Conference on Product Development and Manufacturing Technology, University of Strathclyde, April 1971.

Cutkosky, M.R. and Tenenbaum, J.M., "A methodology and computational framework for concurrent product and process design," Stanford University, Mechanical Engineering Department.

Cutkosky, M.R., Tenenbaum, J.M. and Muller, D., "Features in process-based design," *Computers in Mechanical Engineering*, ASME, July 1988, pp. 1–13.

Cutkosky, M.R., Tenenbaum, J.M. and Muller, D., "Features in process-based design," ASME Conference of Computers in Engineering, July 31–August 4, 1988, pp. 557–562.

Davies, B.J., "Application of expert systems in process planning," *Annals of the CIRP*, vol. 35, no. 2, 1986.

Davies, B.J. et al., "The Integration of process planning with CAD CAM including the use of expert systems," Proceedings of the International Conference on CAPE, April 1986.

Davies, B.J., "Expert systems in process planning," 7th International Conference on the Computer as a Design Tool, 2–5 September, 1986.

Davies, B.J. and Darbyshire, I.L., "The use of expert systems in process planning," *Annals of the CIRP*, vol. 33, no. 1, 1984, pp. 303–306.

Davies, B.J., Darbyshire, I.L., Wright, A.J. and Zhang, K.F., "IKBS process planning system for rotational parts," Intelligent Manufacturing System, II., 2nd International Summer Seminar, (Milacic, V.R. ed.), August 24–29, 1987, pp. 27–38.

Davies, B.J., "Knowledge-based systems in production engineering," *Annals of the CIRP*, vol. 35, no. 2, 1986.

Descotte, Y. and Latombe, J.C., "GARI-A problem solver that plans how to machine mechanical parts," Proceedings of 7th International Joint Conference on Artificial Intelligent, August 1981, pp. 329–347.

Descotte, Y. and Latombe, J.C., "GARI: an expert system for process planning," *Solid Modeling by Computers: From Theory to Applications*, M. Picketts and J.W. Barpe (eds.), Plenum Press, 1984, pp. 329–347.

DeVor, R.E., Zdeblick, W.J., Tipnis, V.A. and Buescher, S., "Development of mathematical models for process planning of machining operations," Proceedings of Sixth NAMRC, 1978, pp. 395–401.

Du, P. and Liu, J., "The use of expert system in computer aided process planning, Proceedings of the 7th PROLAMAT Conference, June 14–17, 1988.

Dunn, M.S. and Mann, W.S., "Computerized production process planning", Proceedings of the 15th Numerical Control Society Annual Meeting and Technical conference, 1978.

Dunn, M.S., "Computerized production process planning for machined cylinderical parts," 19th NCS Annual Proceedings, 1982, pp. 162–173.

El-Midany, T.T. and Davies, B.J., "AUTOCAP-a dialogue system for planning the sequence of operations for turning parts," *International Journal of Machine Tool Design*, vol. 21, no. 3/4, 1981, pp. 175–191.

Englert, P.J. and Wright, P.K., "Principles for part setup and workholding in automated manufacturing," *Journal of Manufacturing Systems*, vol. 7, no. 2, 1988, pp. 147–161.

Eshel, G. and Barash, M.M., "Automatic generation of multi-technology process-outlines," PED-Vol 21, *Integrated and Intelligent Manufacturing*, (eds Liu, C.R. and Chang, T.C.), ASME, WAM, Anaheim, Calif., December 7–12, 1986, pp. 193–218.

Eshel, G., "Automatic generation of process outlines of forming and machining processes," Ph.D. Thesis, Purdue University, August, 1986.

Eshel, G., Barash, M., and Chang, T.C., "A rule-based system for automatic generation of process outlines of deep-drawng processes," Automated Process Planning Systems, ASME Winter Annual Meeting, Miami Beach, Fla., Refereed Bound Volume, November 18–22, 1985, pp. 1–18.

Eshel, G., Barash, M., and Chang, T.C., "A rule-based system for automatic generation of process outlines of deep-drawing processes," *Journal of Engineering for Industry*, (in press).

Eshel, G., Barash, M., and Chang, T.C., "Generate & test and rectify—a plan synthesis tactic for automatic process planning," *International Journal for Artificial Intelligence in Engineering*, vol. 3, no. 1, January 1988, pp. 18–34.

Eskicioglu H., "An interactive process planning system for prisimatic parts (ICAPP)," *CIRP Annals*, vol. 2, no. 1, 1983, pp. 365–370.

Eskicioglu, H. and Davies, B.J., "An interactive process planning system for prismatic parts (ICAPP)," *Annals of the CIRP*, vol. 32, no. 1, 1983.

Eskicioglu, H. and Davies, B.J., "An interactive process planning system for prismatic parts (ICAPP)," *Int. J. of Machine Tool Design & Research*, vol. 21, no. 3/4, 1981, pp. 193–206.

Evans, S. and Sackett, P.J., "Realised performance in a novel approach to computer aided process planning," International conference on Computer Aided Engineering, December 10–12, 1984.

Eversheim, W. and Esch, H., "Automated generation of process plans of prismatic parts," *Annals of CIRP*, vol. 32, no. 1, 1983, pp. 361–364.

Eversheim, W. and Fuchs, H., "Integrated generation of drawing, process plans and NC tapes," *Advanced Manufacturing Technology*, (ed. P. Blake) North-Holland Publishing Company, 1980, pp. 303–314.

Eversheim, W. and Schulz, J., "DISAP—a dialogue oriented process planning system," 1st CIRP Working Seminar on Computer Aided Process Planning, Paris, January 1985.

Eversheim, W. and Schulz, J., "Strategies of process selection for different applications of computer aided process planning," PED-Vol 19, *Computer-Aided/Intelligent Process Planning* (eds. Liu, C.R., Chang, T.C. and Komanduri, R.), ASME WAM, Miami Beach, Fla., November 17–22, 1985, pp. 55–64.

Eversheim, W., Diels, A., and Rozenfeld, H., "Chaning requirements for CAP-systems lead to a new CAP data model," Proceedings of the 19th CIRP International Seminar on Manufacturing Systems, Pennsylvania State University, June 1–2, 1987, pp. 9–15.

Eversheim, W., et al., "Application of automatic process planning and NC-programming," Proceedings of Autofact West, Society of Manufacturing Engineers, November 17–20, 1980.

Eversheim, W., Holz, B. and Zons, K.H., "Application of automatic process planning and NC programming," Proceedings of Autofact West, Society of Manufacturing Engineers, November 17–20, 1980, pp. 779–800.

Fainguclernt, D., et al., "Computer aided tolerancing and dimensioning in process planning," *Annals of the CIRP*, vol. 35, no. 1, 1986.

Fujita, S., et al., "Study of practical computer aided process planning based on expert systems," Proceedings of 7th PROLAMAT Conference, June 14–17, 1988.

Gams, M., Cestnik, B., Sluga, L. and Butala, P., "OPEX—an expert system for CAPP," 6th International Workshop in Expert Systems and Their Applications, vol. 2, April 1986, pp. 1251–1261.

Giusti, F., Santochi, M., and Dinia, G., "COATS: an expert system for optimal tool selection," *Annals of CIRP*, vol. 35, no. 1, 1986, pp. 337–340.

Halevi, G. and Stout, K.J., "A computerised planning procedure for machined components," *Production Engineer*, April 1977, pp. 37–42.

Halevi, G. and Weill, R., "Development of flexible optimum process planning procedures," *Annals of the CIRP*, vol. 29, no. 1, 1980, pp. 313–317.

Hancock, T.M., "Integration of design, planning, and manufacturing subystems in sheet metal processing," 1988 International Conference on Computer Integrated Manufacturing, May 23–25, 1988, pp. 138–143.

Hannam, R G. and Plummer, J.C.S., "Capturing production engineering practice within a CADCAM system," *International Journal of Production Research*, vol. 22, no. 2, 1984, pp. 267–280.

Hayes, C. and Wright, P., "Automated planning in the machining domain," Symposium on Knowledge-Based Expert System for manufacturing, The Winter Annual Meeting of the ASME, December 7–12, 1986, pp. 221–232.

Hayes, C.C. and Wright, P.K., "Setup planning in machining: an expert system approach," 1989 NSF Conference on Advances in Manufacturing System Integration & Process, January 1989, pp. 441–443.

Hayes, C. and Wright, P., "Automating process planning: using feature interactions to guide search," *Journal of Manufacturing Systems*, vol. 8, no. 1, 1989, pp. 1–14.

Hinderman, J., "Composites GPP system update," Proceedings of the 1988 DCLASS Conference, March 7–11, 1988.

Hinduja, S. and Kroeze, B., "Selection of tools for the finishing operations on turned components," Proceedings of the PROLAMAT 85 Conference, 1985.

Hoffmann, P. and Garzo, A., "The process planning system of Lang Engineering Works," The 5th International IFIP/IFAC Conference on Programming Research and Operations, May 1982.

Hon, K.K. and Ismail, H., "A knowledge based system for the selection of hole-making processes," 9th International Conference on Production Research, August 17–20, 1987.

Horvath, M., "Semi-generative process planning for part manufacturing," SME Technical Paper Series, MS79-153, 1979.

Huang, N.K., et al., "CAOS-a generative computer aided operation scheming system for single spindle automatic lathe," PED-Vol 21, *Integrated and Intelligent Manufacturing*, (eds. Liu, C.R. and Chang, T.C.), ASME, WAM, Anaheim, Calif., December 7–12, 1986, pp. 1227–236.

Hummel, K.E. and Brooks, S.L. "Using hierarchically structured problem-solving knowledge in a rule-based process planning system," Bound Volume, ASME WAM, 1987.

Hummel, K.E. and Brooks, S.L., "Symbolic representation of manufacturing features for an automated process planning system," Bound Volume of the Symposium on Knowledge Based Expert Systems for Manufacturing, WAM ASME, Dec. 7–12, 1986, pp. 233–243.

Hummel, K.E., "An expert machine tool planner," ASME Computers in Engineering Conference, 1987.

Inui, M., Kinosada, A., Suzuki, Kimura, and Sata, T., "Automatic process planning for sheet metal parts with bending simulation,"

PED Vol 25, *Intelligent and Integrated Manufacturing Analysis and Synthesis*, Bound Volume of ASME-WAM. 1987.

Inui, M., Suzuki, H., Kimura, F. and Sata T., "Extending process planning capabilities with dynamic manipulation of product models," Proceedings of the 19th CIRP International Seminar on Manufacturing Systems, Pennsylvania State University, June 1–2, 1987, 273–280.

Irizarry-Gaskins, V. and Chang, T.C., "Knowledge based process planning for electronic assembly," *Journal of Intelligent and Robotic Systems*, (in press).

Irizarry-Gaskins, V.M. and Chang, T.C., "A process planning methodology in electronic assembly," Annual Conference of the Society for Integrated Manufacturing (SIM), November 12–15, 1989, pp. 919–926.

Irizarry-Lopez, V.M., "A methodology for the automatic generation of process plans in an electronic assembly environment," Ph.D. Thesis, Purdue University, August 1989.

Irizarry-Lopez, V.M., Chang, T.C., "Knowledge based process planning for electronic assembly," second international symposium Robotics and Manufacturing Research, Education and Applications, November 16–18, 1988.

Iwata, K. and Fukuda, Y., "KAPPS: Know-how and knowledge assisted production planning system in the machining shop," Proceedings of the 19th CIRP International Seminar on Manufacturing Systems, Pennsylvania State University, June 1–2, 1987.

Iwata, K. and Sugimura, N., "A knowledge based computer aided process planning system for machining parts," Sixteenth CIRP International Seminar on Manufacturing Systems, Tokyo, 1984.

Iwata, K. and Sugimura, N., "An integrated CAD/CAPP system with know-hows on machining accuracies of Parts," PED-Vol 19, *Computer-Aided/Intelligent Process Planning* (eds. Liu, C.R., Chang, T.C., and Komanduri, R.), ASME WAM, Miami Beach, Fla., November 17–22, 1985, pp. 121–130.

Iwata, K. and Sugimura, N., "An integrated CAD/CAPP system with know-hows on machining accuracies of parts," *Journal of Engineering for Industry*, Transactions of the ASME, ASME, May 1987, pp. 128–133.

Iwata, K., and Fukuda, Y., "Representation of know-how and its application of machining reference surface in computer aided process planning," *Annals of the CIRP*, vol. 35, no. 1, 1986.

Iwata, K., Kakino, Y., Oba, F., and Sugimura, N., "Development of non-part family type computer aided production planning system CIMS/PRO," *Advanced Manufacturing Technology*, (ed. P. Blake) North-Holland Publishing Company, 1980, pp. 171–184.

Iwata, K., Kakino, Y., Oba, F., and Sugimura, N., "Development of non-part family type computer aided production planning system CIMS/PRO," Fourth International IFIP/IFAC Conference on PRO-LAMAT 79, (ed. P. Blake), North-Holland Publishing Company, May 21–23, 1979, pp. 171–185.

Iwata, K., "Knowledge based computer aided process planning," Intelligent Manufacturing System, II., 2nd International Summer Seminar, (Milacic, V.R. ed.), August 24–29, 1987, pp. 3–25.

Jha, N.K., "Optimum process planning (OPIPP) and expert system process planning (ESYPP): A key to CAD/CAM integration," Proceedings of the 1988 ASME International Computers in Engineering Conference and Exhibition, ASME, July 31–August 4, 1988, pp. 451–458.

Joneja, A. and Chang, T.C., "A generalized framework for automatic planning of fixture configuration," ASME Winter Annual Meeting, December 10–15, 1989.

Jorgensen, J., "Building expert systems with DCLASS," DCLASS Users meeting, Brigham Young University, 1985.

Jorgensen, J., "Computer integrated manufacturing and expert systems," ATV-SEMAPP Workshop. 1986.

Joshi, S., Chang, T.C., and Liu, C.R., "An automated process planning system structure based on AI," Proceedings ASME International Conference on Computers in Engineering, vol. 1, ASME, July 20–24, 1986, pp. 247–254.

Joshi, S., Chang, T.C., and Liu, C.R., "Process planning formalization in an AI framework," *International Journal of Artificial Intelligence in Engineering*, vol. 1, no. 1, 1986, pp. 45–53.

Joshi, S., Vissa, N., and Chang, T.C., "Expert process planning system with solid model interface," *Expert Systems: Design and Management of Manufacturing Systems*, (A Kusiak ed.), Taylor & Francis, 1988, pp. 111–136.

Joshi, S., Vissa, N., and Chang, T.C., "Expert process planning system with solid model interface," *International Journal of Production Research*, vol. 26, no. 5, 1988, pp. 863–885.

Jun, J.S. and McLean, C.R., "Control of an automated machining workstation," *IEEE Control System Magazine*, vol. 8, no. 1, February 1988, pp. 26–30.

Kakino, Y., Ohba, F., Moriwaki, T., and Iwata, K., "A new method of parts description for computer aided process planning," *Advances in Computer-Aided Manufacturing,* (ed. D. McPherson) North-Holland Publishing Co., 1977, pp. 197–213.

Kanumury, M., "AMPS—An automatic manufacturing process planning system," M.S. thesis, School of Industrial Engineering, Purdue University, 1988.

Kanumury, M., Shah, J., and Chang, T.C. "An automatic process planning system for a quick turnaround cell—An integrated CAD and CAM system," USA-Japan Symposium on Flexible Automation, ASME, July 18th, 1988.

Kashyap, R.L., Smit, H.J., Tsatsoulis, C., and Wiggins, L.K., "An intelligent system for integrating process planning and design," Proceedings of the 1988 IEEE International Conference on Robotics and Automation, April 24–29, 1988, pp. 1297–1299.

Kimura, F. and Sata, T., "Integration of design and manufacturing activities based on object modeling," Advances in CAD/CAM, North-Holland Publishing Company, 1983.

Kishinami, T., et al. "An integrated approach to CAD/CAPP/CAM based on cell-constructed-geometric-model (CCM)," 3,(2), 215–220. *Robotics and Computer-Integrated Manufacturing,* vol. 3, no. 2, 1987, pp. 215–220.

Koenig, D.T., "Computer aided process planning for computer integrated manufacturing," Proceedings of the 19th CIRP International Seminar on Manufacturing Systems, Pennsylvania State University, June 1–2, 1987.

Koloc, J., "Miturn, A computer-aided production planning system for numerically controlled lathes," Proceedings of the Second International Conference on Product Development and Manufacturing Technology, University of Strathclyde, April 1971.

Kramer, T., "The design protocol, part design editor, and geometry library of the vertical workstation of the automated manufacturing research facility at the national bureau of standards," Document No. NBSIR 88-3731, National Bureau of Standards, 1988.

Kramer, T.R. and Jun, J.S., "Software for an automated machining work station," NBS report, July 1986.

Kramer, T.R. and Jun, J.S., "Software for an automated machining work station," Proceedings of the 1986 International Machine Tool Technical Conference, Chicago, Ill. September 1986, pp. 12–9 to 12–44.

Kramer, T.R., "Process plan expression, generation, and enhancement for the vertical workstation milling machine in the automated manufacturing research facility at the national bureau of standard," National Bureau of Standards, 1987, (51 pages).

Krause, F.L., Armbrust, P., and Bienert, M., "Methodbases and product models as bases for integrated design and manufacturing," *Robotics and Computer Integrated Manufacturing*, vol. 4, no. 1–2, 1988, pp. 33–40.

Kung, H., "An investigation into development of process plans from solid geometric modeling representation." Ph.D. thesis, Oklahoma State University, Okla., 1984.

Kusiak, A., "Automated process planning in flexible manufacturing systems," Proceedings of the International Conference on Advances in Manufacturing, IFS, Ltd and North-Holland, October 9–11, 1984, pp. 61–69.

Kyttner, R., et al., "Framework for integrated computer aided process planning and scheduling systems," Proceedings of the 7th PROLAMAT Conference, Dresden, GDR, June 14–17, 1988.

Lagoude, Y. and Tsang, J.P., "A plan representation structure for expert planning systems," Bound Volume, Symposium on Computer-Aided/Intelligent Process Planning, ASME, WAM, Miami Beach, Fla., November 1985.

Lambourne, E.B., "Towards integration of computer-aided design, manufacture and production management," *Computer-Aided Engineering Journal*, December 1986.

Latombe, J-C. and Dunn, M.S., "XPS-E: an expert system for process planning, man or machine –A choice of intelligence," Proceedings of CAM-I's 13th Annual Meeting and Technical Conference, 1984, pp. 3.7-3.15.

Lenau, T. and Alting, L., "Artificial intelligence for process selection," *Journal of Mechanical Working Technology*, vol. 17, 1988, pp. 33–49.

Lenau, T. and Alting, L., "XPLAN-an expert process planning system," Second International Expert Systems Conference, London, UK, 30 September–2 October 1986.

Li, J., Han, C. and Ham, I., "CORE-CAPP a company-oriented semi-generative computer automated process planning system," Proceedings of the 19th CIRP International Seminar On Manufacturing Systems, June 1987, pp. 219–225.

Li, R.K. and Bedworth, D.D., "A framework for the integration of computer-aided design and computer-aided process planning," *Computer and Industrial Engineering*, vol. 14, no. 4, 1988, pp. 395–413.

Lin, L. and Bedworth, D.D., "A semi-generative approach to computer-aided process planning using group technology," *Computers and Industry Engineering,* vol. 14, no. 2, 1988, pp. 127–137.

Liou, M. and Sheu, P., "Automatic process pre-planning in manufacturing environments," 1988 pp. 1294–1296.

Liu, D., "Utilization of artificial intelligence in manufacturing," Proceedings of Autofact 6 Conference, October 1–4, 1984, pp. 2.60–2.78.

Liu, Y.S. and Allen, R., "A proposed synthetic, interactive process planning system," Proceedings of the International Conference on CAPE, Edinburgh, UK, April 1986, pp. 201–207.

Lu, S.C-Y. and Subramanyam, S., "A computer-based environment for simultaneous product and process design," Bound Volume, Symposium on Advances in Manufacturing System Engineering, ASME, WAM, November 27–December 2, 1988, pp. 35–46.

Major, F. and Grottke, W., "Knowledge engineering within integrated process planning systems," *Robotics and Computer-Integrated Manufacturing,* vol. 3, no. 2, 1987.

Matsushima, K., Okada, N., and Sata, T., "The integration of CAD and CAM by application of artificial intelligence," *Annals of the CIRP,* vol. 31, no. 1, 1982, pp. 329–332.

Matsushima, K., Katayama, T., Sata, T., Yoshizum, T., and Azumi, K., "Development of a process planning system for the pressure vessels," *CIRP Annals,* vol. 28, no. 1, 1979.

McMullen, R.J. and Thewes, R.A., "COGERS: a practical approach to computer aided manufacturing routing," Tooling and Production, Feb. 1978.

Milacic, R.V. and Kalajdzic, M., "Logical structure of manufacturing process-design-fundamentals of an expert system for manufacturing process planned," Proceedings of the 16th CIRP International Seminar on Manufacturing Systems, 1984.

Milacic, R.V., "SAPT-expert system for manufacturing process planning," Ped. Vol. 19, ASME Winter Annual Meeting 1985.

Milacic, V.R., Urosevic, M., Veljovic, A., Miler, A., and Race, I., "SAPT-expert system based on hybrid concept of group technology," Proceedings of the 19th CIRP International Seminar on Manufacturing Systems, Pennsylvania State University, June 1–2, 1987, pp. 189–195.

Milacic, V.R., Urosevic, M., Veljovic, A., Miler, A., and Race, I., "SAPT-expert system based on hybrid concept of group technology," Intelligent Manufacturing System, II., 2nd International Summer Seminar, (Milacic, V.R. ed.), August 24–29, 1987, pp. 39–51.

Mittal, R.O. and Lewis, R.L., "A 'Micro' process planning system based on integer programming for prismatic parts produced on horizontal machining centers," *Annals of Operations Research*, vol. 17, no. 1–4, January 1989, pp. 273–290.

Mouleeswaran, C.B., "PROPLAN—A knowledge based expert system for manufacturing process planning," MS thesis, University of Chicago, Illinois, 1984.

Mouleeswaran, C.B. and Fischer, H.G., "A knowledge based environment for process planning," Proceedings of the 1st International Conference on the Application of Artificial Intelligence in Engineering Problems, April 1986, pp. 1013–1028.

Nakazawa, H. and Suh, N.P., "Process planning based on information concept," Unpublished Paper, MIT, Mass., 1982.

Nau, D.S. and Luce, M., "Knowledge representation and reasoning techniques for process planning: Extending SIPS to do tool section," Proceedings of the 19th CIRP International Seminar on Manufacturing Systems, Pennsylvania State University, June 1–2, 1987.

Nau, D.S., "SIPP Reference Manual," Technical Report, Computer Science Department, University of Maryland, 1985.

Nau, D.A. and Gray, M., "SIPS: an Application of hierarchical knowledge clustering to process planning," PED-Vol 21, *Integrated and Intelligent Manufacturing*, (eds Liu, C.R. and Chang, T.C.), ASME, WAM, Anaheim, Calif., December 7–12, 1986, pp. 219–225.

Nau, D.S. and Chang, T.C., "A knowledge based approach to process planning," PED-Vol 19, *Computer-Aided/Intelligent Process Planning* (eds Liu, C.R., Chang, T.C., and Komanduri, R.), ASME WAM, Miami Beach, Fla., November 17–22, 1985, pp. 65–72.

Nau, D.S. and Chang, T.C., "Prospects for process selection using AI," *Computers in Industry*, vol. 4, no. 3, 1983, pp. 253–263.

Nau, D.S. and Chang, T.C., "Prospects for process selection using artificial intelligence," *Computer-Aided Process Planning*, J. Tulkoff, (ed.), Society of Manufacturing Engineers, 1985.

Nau, D.S., "Hierarchical abstraction for process planning," Second International Conference on Applications of Artificial Intelligence in Engineering, 1987.

Niebel, B. W., "Mechanized process selection for planning new designs," ASME paper, No. 737, 1965.

Okino, N., and Kubo, H., "Technical information system for computer-aided design, drawing, and manufacturing," Proceedings of the Second PROLAMAT 73, 1973.

Olson, W. W. and Devries, W. R., "Logical basis for process planning," Proceedings of the 19th CIRP International Seminar on Manufacturing Systems, Pennsylvania State University, June 1–2, 1987.

Pande, S.S. and Palsule, N.H., "PC-CAPPS—a computer-assisted generative process planning system for turned components," *Computer Aided Engineering*, vol. 5, no. 4, August 1988, pp. 163–168.

Park, M. W. and Davies, B. J., "Integration of process planning into CAD using Iges," Proceedings of the 19th CIRP International Seminar on Manufacturing Systems, Pennsylvania State University, June 1–2, 1987.

Pellegrino, S.A. and Ham, I., "Computer aided milling operation planning Program—CAMOPP," 11th NAMRC, 1983, pp. 484–489.

Phillips, R., Arunthavanathan, V., and Zhou, X.D., "An intelligent design and process planning system," Proceedings of the 19th CIRP International Seminar on Manufacturing Systems, Pennsylvania State University, June 1–2, 1987, pp. 17–22.

Phillips, R.H., Zhou, X.D., and Mouleeswaran, C., "MICROPLAN: a microcomputer based expert system for generative process planning," Symposium on Knowledge-Based Expert System for Manufacturing, The Winter Annual Meeting of the American Society of Mechanical Engineers, Anaheim, Calif., December 7–12, 1986, pp. 263–273.

Pinte, J. and Carlier, J., "Process planning logic extraction study," CAM-I Publication R-84-PPP-02, 1984.

Preiss, K. and Shai, O., "Process planning by logic programming," *Robotics and Computer-Integrated Manufacturing*, vol. 5, no. 1, 1989, pp. 1–10.

Rahman, M. and Narayanan, V., "An expert system for process planning," *Robotics and Computer-Integrated Manufacturing*, vol. 3, no. 3, 1987, pp. 365–372.

Richardson, J., "DCLASS applications in production engineering," Proceedings of the 1988 DCLASS Conference, March 7–11, 1988.

Sack, C.F., Jr., "Computer managed process planning—a bridge between CAD and CAM," Proceedings of Autofact 4 Conference, November 30–December 2, 1982.

Sack, Jr., C.F., "CAM-I's experimental planning system, XPS-1," Proceedings of Autofact 5 Conference, November 14–17, 1983.

Sakamoto, C., et al., "POPULAR-an automatic process planning system," Proceedings of the 19th CIRP International Seminar on Manufacturing Systems, Pennsylvania State University, June 1–2, 1987.

Santochi, M. and Giusti, F., "PICAP: a fully integrated package for process planning of rotational parts," 18th CIRP Seminar on Manufacturing Systems, 1986.

Santochi, M., "PICAP," Short description of research activity on CAPP Istituto di Technologia Meccanica di Pisa, Italy, 1st report, Short description of research activity on CAPP, Istituto di Technologia Meccanica di Pisa, Italy, August 1985.

Shaw, M.J., Park, S., and Menon, U., "Incorporating machine learning in knowledge-based process planning systems: an explanation-based approach,"The Beckman Institute for Advanced Science and Technology, University of Illinois at Urbana-Champaign, October 1988.

Sluga, A., Butala, P., Lavrac, N., and Gams, M., "An attempt to implement expert system techniques in CAPP," *Robotics & Computer-Integrated Manufacturing*, vol. 4, no. 1–2, 1988, pp. 77–82.

Smith, R.M., "Computer-aided, fully generative process planning," *Manufacturing Engineering*, May 1981.

Sohal, G.S. and Eyada, O.K., "An expert system for the selection of indexable inserts and tool holders," the Second International Conference on Industrial & Engineering Applications of Artificial Intelligence & Expert Systems, June 6–9, 1989.

Spadoni, M. and Mutel, B., "A new approach to automatic generation of process-planning," Proceedings of the 19th CIRP International Seminar on Manufacturing Systems, Pennsylvania State University, June 1–2, 1987, pp. 121–125.

Spur, G, Krause, F.-L., and Anger, H.-M., "Working techniques of computer-aided process planning," Advanced Course on Computer-Integrated Manufacturing, (CIM 83), Sept 5 to 16, 1983.

Spur, G., Krause, F.-L., and Grottke, W., "Advanced methods for generative process planning," 1st CIRP working seminar on Computer Aided Process Planning, Paris, January 1985.

Spur, G., Seliger, G., Felsing, W., and Hsieh, L.H., "Process planning for sensor-based surface finishing with industrial robots," PED Vol 25, symposium on Intelligent and Integrated Manufacturing Analysis and Synthesis, ASME WAM, Boston, Mass., December 13–18, 1987, pp. 215–227.

Srihari, K. and Greene, T.J., "Alternate routings in CAPP implementation in a FMS," *Computers and Industrial Engineering*, vol. 15, no. 1–4, 1988, pp. 41–50.

Srinivasan, R. and Liu, C.R., "On some important geometric issues in generative process planning," PED Vol 25, *Intelligent and Integrated Manufacturing Analysis and Synthesis,* Bound Volume, ASME Winter Annual Meeting, Boston, Massachusetts, December 13–18, 1987. PP 229–243.

Srinivasan, R., "A generalized analytical methodology for generative process planning," Ph.D. Thesis, Purdue University, West Lafayette, Ind., December 1986.

Ssemakula, M. and Davies, B.J., "Integrated process planning and NC programming for prismatic parts," Proceedings of 1st International Machine Tool Conference, IFS Publication, June 1984, pp. 143.

Strohmeier, A.H., "Implementing computer-aided process planning: Rockwell International case study," *CIM Review*, Fall 1987.

Subramanyam, S. and Lu, S. C-Y., "An integrated AI/OR approach to operation planning based on process performance models," Proceedings of the 16th NAMRC Conference, 1988.

Sundaram, R.M. and Cheng, T.J., "Microcomputer-based process planning using geometric programming," *International Journal of Production Research*, vol. 24, no. 1, 1986, pp. 119–127.

Sundaram, R.M., "Process planning and machining sequence," Proceedings of the 19th CIRP International Seminar on Manufacturing Systems, Pennsylvania State University, June 1–2, 1987.

Takano, K., Kakino, Y., and Swata, K., "CAPP—compurt aided process planning for job shop with Integrated manufacturing system," Proceedings of the International Conference on Production Engineering, Tokyo, Japan, 1975.

Terwilliger, J.P., Process planning for electric assembly," Master's Thesis, Purdue University, West Lafayette, Indiana, USA, August 1985.

TIPNIS, Inc., EXPERT Manufacturing Planning System, 1986, TIPNIS Inc., Brochures and information material, 10815 Indeco Drive, Cincinnati, Ohio, USA.

Tonshoff, H.K. and Hellberg, K., "Knowledge engineering for automatic generative process planning," International Conference on Advanced Manufacturing Systems and Technology, AMST-87.

Tonshoff, H.K., Beckendorff, U., and Schaele, M., "Some approaches to represent the interdependence of process planning and process Control," Proceedings of the 19th CIRP International Seminar on Manufacturing Systems, Pennsylvania State University, June 1–2, 1987, pp. 257–271.

Tou, J.T., "Design of expert systems for integrated production automation," *Journal of Manufacturing Systems*, vol. 4, no. 2, 1985.

Tsang, J., "The Propel process planning," Proceedings of the 19th CIRP International Seminar on Manufacturing Systems, Pennsylvania State University, June 1–2, 1987, pp. 71–77.

Tsang, J.P., "Genericity in process planning systems," 1st International conference on Applications of AI to Engineering Problems, Southampton, UK, April 1986.

Tsatsoulis, C. and Kashyap, R.L., "A system for knowledge-based process planning," *Artificial Intelligence in Engineering*, vol. 3, no. 2, 1988, pp. 61–75.

Tsatsoulis, C., "Using dynamic memory structures in planning and its applications to manufacturing," Ph.D. Thesis, Purdue University, West Lafayette, Ind., December 1987.

Tulkoff, J. "GT-based generative process planning," SME Technical Report, MS83-915, 1983.

Tulkoff, J. "Lockheed's GENPLAN," 18th NCS Annual Procceedings, May 1981.

Tulkoff, J., "Lockheed's GENPLAN," in Proceedings of 18th Numerical Control Society Annual Meeting and Technical Conference, Dallas, Tex., 1981, pp. 147–421.

Uemura, N., Yokoi, S., Hisatomi, Y., and Inagaki, K., "Development of a Process Planning System Using Knowledge Engineering and Geometric Processing," *NEC Research Development*, no. 91, October 1988, pp. 111–115.

van Houten, F.J.A. and Tiemersma, J.J., "An Adaptive Control Module For Round."

van Houten, F.J.A.M. and Kals, H.J.J., "ROUND a flexible technology based process and operations planning system for NC lathes," Proceedings of the 16th CIRP International seminar on Manufacturing Systems, Tokyo, 1984.

van Houten, F.J.A.M., van't Erve, A.H. and Kals, H.J.J., "PART a feature based CAPP System," CIRP Manufacturing Systems, June 1989.

van Houten, M. F.J., "Strategy in generative planning of turning processes," *Annals of CIRP*, vol. 35, no. 1, 1986, pp. 331–335.

van't Erve, A.H. and Kals, H.J.J., "XPLANE, a generative computer aided process planning system for part manufacturing," *Annals of CIRP*, vol. 35, no. 1, 1986, pp. 325–329.

van't Erve, A.H. and Kals, H.J.J., "XPLANE, a knowledge-based driven process planning expert system," Proceedings of CAPE, April 1986, pp. 41–46.

van't Erve, A.H. and Kals H.J.J., "XPLANE, a Generative Computer aided process planning system for part manufacturing," *Annals of CIRP*, vol. 35, 1986, pp. 324–329.

Vogel, S.A. and Dawson, D., "Integrated process planning at General Electric's Aircraft Engine Group," Proceedings of Autofact West, Society of Manufacturing Engineers, November 17–20, 1980.

Vogel, S.A. and Adard, E.J., "The AUTOPLAN process planning system," 18th NCS Annual Proceedings, 1981.

Wang, C.-L., Bagchi, A., and Ahluwalia, R.A., "DMAP: a computer integrated system for design and manufacture of axisymmetric parts," Bound Volume, *Knowledge-Based Expert Systems for Manufacturing*, ASME, Winter Annual Meeting, Anaheim, CA, Dec 7–12, 1986, pp. 327–338.

Wang, H.-P. and Wysk, R.A., "A knowledge-based approach for automated process planning," *International Journal of Production Research*, vol. 26, no. 6, 1988. pp. 999–1014.

Wang, H.-P. and Wysk, R.A., "A knowledge-based computer aided process planning system," Proceedings of the 19th CIRP International Seminar on Manufacturing Systems, Pennsylvania State University, June 1–2, 1987.

Wang, H.-P. and Wysk, R.A., "Intelligent reasoning for process planning," *Computers in Industry*, vol. 8, 1987.

Wang, H.-P., "A layered architecture for manufacturing operation planning," *Computers and Industrial Engineering*, vol. 14, no. 2, 1988, pp. 201–210.

Wang, H.P. and Wysk, R.A., "Micro-GEPPS: A microcomputer based process planning system," PED-Vol 19, *Computer-Aided/Intelligent Process Planning* (eds Liu, C.R., Chang, T.C., and Komanduri, R.),

ASME WAM, Miami Beach, Fla., November 17–22, 1985, pp. 139–150.

Wang, H.P. and Wysk, R.A., "AIMSI: a prelude to a new generation of integration CAD/CAM systems," *International Journal of Production Research*, vol. 26, no. 1, 1988, pp. 119–131.

Wang, H.P. and Wysk, R.A., "Intelligent reasoning for process planning," Technical Paper, Department of Industrial and Systems Engineering, Penn State University, 1986.

Wang, H.P. and Wysk, R.A., "Intelligent reasoning for process planning," *Computers in Industry*, vol. 8, 1987, pp. 293–309.

Wang, H.P. and Wysk, R.A., "Micro-GEPPS—a microcomputer based process planning system," Bound Volume, Symposium on Computer-Aided/Intelligent Process Planning, ASME Winter Annual Meeting, Miami Beach, FL, pp. 139–150, November 17–22, 1985.

Wang, H.P. and Wysk, R.A., "Turbo-CAPP: A knowledge-based computer-aided process planning," Proceedings of the 19th CIRP International Seminar on Manufacturing Systems, Pennsylvania State University, June 1–2, 1987, pp. 161–167.

Wang, H.P., "Intelligent reasoning for process planning," Ph.D. Thesis, Pennsylvania State University, State College, 1986.

Weill, R., "Integrating dimensioning and tolerancing in computer-aided process planning," *Robotics and Computer Integrated Manufacturing*, vol. 4, no. 1–2, 1988, pp. 41–48.

Willis, D., Donaldwon, I.A., Ramage, A.D., Murray, J.L., and Williams, M.H., "A knowledge-based system for process planning based on a solid modeller," *Computer-Aided Engineering Journal*, vol. 6, no. 1, February 1989, pp. 21–26.

Wong, C.L., Bagchi, A., and Ahluwalia, R.A., "DMAP: A computer integrated system for design and manufacturing of axisymmetric parts," Bound Volume, *Knowledge-Based Expert Systems for Manufacturing*, ASME WAM, 1986, pp. 327–338.

Wong, C.L., "DMCP: A computer assisted system for the design and manufacturing of cylindrical parts," MS Thesis, the Ohio State University, 1985.

Woodhead, R., Dobolyi, Z., and De Pennington, A.D., "Process planning as an application for expert systems technology," PED-Vol 21, *Integrated and Intelligent Manufacturing*, (eds Liu, C.R. and Chang, T.C.), ASME, WAM, Anaheim, Calif., December 7–12, 1986, pp. 143–156.

Wright, A.J., et al., "EXCAP and ICAPP: Integrated knowledge based systems for process-planning components," Proceedings of the 19th CIRP International Seminar on Manufacturing Systems, Pennsylvania State University, June 1–2, 1987.

Wu, M.C., "A new methodology for automatic process planning and execution based on adaptive information modeling," Ph.D. Thesis, Purdue University, W. Lafayette, Ind., 1988.

Wysk, R.A., "An automated process planning and selection program: APPAS," Ph.D. Thesis, Purdue University, West Lafayette, Ind.

Wysk, R.A., Barash, M.M., and Moodie, C.M., "Unit machining operations: an automated process planning and selection program," *Journal of Engineering for Industry*, Transactions of the ASME, vol. 102, November 1980. pp. 298–302.

Wysk, R.A., Chang, T.C., and Davis, R.P., "Analytical Techniques in Automated Process Planning," MAPEC the NSF Manufacturing Productivity Education Committee, Educational Module, Purdue University, January 1980. (59 pages)

Zhang, H.C. and Alting, L., "Introduction to an intelligent process planning system for rotational parts," Proceedings of the Advances in Manufacturing Systems Engineering Symposium, ASME-WAM 1988, Chicago, Illinois, USA, 28 November–2 December, 1988.

Zhang, H.C., "Overview of CAPP and development of XPLAN," Institute of Manufacturing Engineering, The Technical University of Denmark, Publication No. 87-26. 1987.

Zhang, H.C. and Alting, L., "XPLAN-R: an expert process planning system for rotational components," Proceedings of IEE 1988 Integrated Systems Conference, St. Louis, Missouri, USA, October 30–November 2, 1988.

A.4 Expert Systems Based Process Planning

Alder, G.M., "Selection of machining process by intelligent knowledge based systems," International Conference on Computer Aided Production Engineering, Edinburgh, MEP Publication, 1986, pp. 61–66.

Alting, L., et al., "XPLAN-an expert process planning system and its further development," 27th International MATADOR Conference, April 20–21, 1988.

Barkocy, B.E. and Zdeblick, W.J., "A knowledge-based system for machining operation planning," Proceedings of Autofact 6 Conference, October 1–4, 1984.

Barkocy, B.E. and Zdeblick, W.J., "A knowledge based system for machining operations planning," SME Technical Paper MS 84-716, 1984.

Berenji, H.R. and Khoshnevis, B., "Use of artificial intelligence in automated process planning," *Computers in Mechanical Engineering*, September, 1986, pp. 47–55.

Brooks, S.L., "Applying artificial intelligence techniques to generative process planning systems," M.S. thesis, University of Kansas, May 1986.

Brown, P.F. and McLean, C.R., "Interactive process planning in the AMRF," Knowledge-Based Expert Systems for Manufacturing, ASME Winter Annual Meeting, ASME, December 1986, pp. 245–262.

Chang, T.C. and Terwilliger, J., "A rule based system for printed wiring assembly process planning," *International Journal of Production Research*, vol. 25, no. 10, 1987, pp. 1465–1482.

Chang, T.C. and Terwilliger, J., "PWA–planner—a rule based system for printed wiring assemblies process planning," Proceedings, 9th Annual Conference on Computers and Industrial Engineering, March 18–21, 1987.

Chang, T.C. and Terwilliger, J., "Rule based approach for printed wiring assembly process planning," Proceedings, 8th International Conference on Assembly Automation (8 ICAA), March 31–April 2, 1987.

Chang, T.C., Anderson, D.C., and Mitchell, O.R., "QTC—an integrated design/manufacturing/vision inspection system for prismatic part," Proceedings of the ASME 1988 Computers in Engineering Conference, Vol. 1, July 31–August 3, 1988, pp. 417–426.

Chang, T.C., Anderson, D.C., and O.R. Mitchell, "A short course for the design of a quick turnaround cell," Engineering Research Center for Intelligent Manufacturing Systems, July 26, 1988.

Chang, T.C., "Just in time information generation—a study of automatic process planning," 15th NSF Grantees Conference on Production, research, and Technology, January 8–13, 1989.

Cincinnati Milacron Marketing Co., "Intelligent machining workstation initiative," Program Abstract, prepared under Air Force Contract, Contract No. F33615-86-C-5038, January 1989.

Cutkosky, M.R., Tenenbaum, J.M., and Muller, D., "Features in process-based design," ASME Conference of Computers in Engineering, July 31–August 4, 1988, pp. 557–562.

Davies, B.J., "Application of expert systems in process planning," *Annals of the CIRP*, vol. 35, no. 2, 1986.

Davies, B.J., "Expert systems in process planning," 7th International Conference on the Computer as a Design Tool, 2–5 September, 1986.

Descotte, Y. and Latombe, J.-C., "GARI-A problem solver that plans how to machine mechanical parts," Proceedings of 7th International Joint conference on Artificial Intelligent, August 1981, pp. 329–347.

Du, P. and Liu, J., "The use of expert system in computer aided process planning," Proceedings of the 7th PROLAMAT Conference, June 14–17, 1988.

Eshel, G., "Automatic generation of process outlines of forming and machining processes," Ph.D. Thesis, Purdue University, W. Lafayette, Ind., August 1986.

Eshel, G., Barash, M., and Fu, K.S., "Generating the inclusive test-rule in a rule based system for process planning," PED-Vol 19, *Computer-Aided/Intelligent Process Planning* (eds. Liu, C.R., Chang, T.C., and Komanduri, R.), ASME WAM, Miami Beach, Fla., November 17–22, 1985, pp. 151–166.

Eshel, G., Barash, M., and Chang, T.C., "Generate & Test and Rectify—a plan synthesis tactic for automatic process planning," *International Journal for Artificial Intelligence in Engineering*, vol. 3, no. 1, January 1988, pp. 18–34.

Gams, M., Cestnik, B., Sluga, L., and Butala, P., "OPEX—an expert system for CAPP," 6th International Workshop in Expert Systems and Their Applications, vol. 2, April 1986, pp. 1251–1261.

Giusti, F., Santochi, M., and Dini, G., "COATS: an expert system for optimal tool selection," *Annals of CIRP*, vol. 35, no. 1, 1986, pp. 337–340.

Gupta, T. and Ghosh, B.K., "A survey of expert systems in manufacturing and process planning," *Computers in Industry*, vol. 11, no. 2, 1988, pp. 195–204.

Hayes, C.C., "Planning in the machining domain: using goal interactions to guide search," M.S. Thesis, Carnegie Mellon University, April 1987.

Hayes, C.C. and Wright, P.K., "Setup planning in machining: an expert system approach," 1989 NSF Conference on Advances in

Manufacturing System Integration & Process, January 1989, pp. 441–443.

Hayes, C. and Wright, P., "Automating process planning: using feature interactions to guide search," *Journal of Manufacturing Systems*, vol. 8, no. 1, 1989, pp. 1–14.

Henderson, M.R., and Chang, G.-J., "FRAPP: automated feature recognition and process planning from solid model data," Proceedings of the 1988 ASME International Computers in Engineering Conference and Exhibition, ASME, July 31–August 4, 1988, pp. 529–536.

Hon, K.K. and Ismail, H., "A knowledge based system for the selection of hole-making processes," 9th International Conference on Production Research, August 17–20, 1987.

Hummel, K. and Brooks, S., "Symbolic representation of manufacturing features for an automated process planning system," Bound Volume, *Knowledge-Based Expert Systems for Manufacturing*, ASME WAM, Anaheim, Calif., December 1986.

Hummel, K., "An expert systems based machine tool planner for a distributed automated process planning system," MS Thesis, University of Kansas, 1985.

Husbands, P., Mill, F., and Warrington, S., "A knowledge based process planning system," 2nd International Conference on Applications of AI in engineering (Planning and Design), p. 439.

Inui, M., Suzuki, H., Kimura, F., and Sata, T., "Generation and verification of process plans using dedicated models of products in computers," Bound Volume, *Knowledge-Based Expert Systems for Manufacturing*, Winter Annual Meeting of the American Society of Mechanical Engineers, December 1986, p. 275.

Iwata, K. and Sugimura, N., "A knowledge based computer aided process planning system for machining parts," Sixteenth CIRP International Seminar on Manufacturing Systems, Tokyo, 1984.

Jha, N.K., "Optimum process planning (OPIPP) and expert system process planning (ESYPP): a key to CAD/CAM integration," Proceedings of the 1988 ASME International Computers in Engineering Conference and Exhibition, ASME, July 31–August 4, 1988, pp. 451–458.

Johnson, B.T., "Expert systems in process planning," Proceedings of 4th European Conference Automated Manufacturing, May 1987, pp. 497–506.

Kanumury, M., Shah, J., and Chang, T.C. "An automatic process planning system for a quick turnaround cell—an integrated CAD and CAM system," USA-Japan Symposium on Flexible Automation, ASME, July 18th, 1988.

Lagoude, Y. and Tsang, J.P., "A plan representation structure for expert planning systems," PED-Vol 19, *Computer-Aided/Intelligent Process Planning* (eds. Liu, C.R., Chang, T.C. and Komanduri, R.), ASME WAM, Miami Beach, Fla., November 17–22, 1985, 19–30.

Latombe, J.C., Tsang, J.P., Sack, C.F., Weisser, P.T., and Dunn, M.S., "XPS-E Phase 2A final report," CAM-I publication R-85-PPP-02, July, 1985. Berenji, H.R., "Use of artificial intelligence in automated process planning," *Computers in Mechanical Engineering*, September 1986.

Latombe, J.C. and Dunn, M.S., "XPS-E: an expert system for process planning, man or machine—a choice of intelligence," Proceedings of CAM-I's 13th Annual Meeting and Technical Conference, 1984, pp. 3.7-3.15.

Lenau, T. and Alting, L., "Artificial intelligence for process selection," *Journal of Mechanical Working Technology*, vol. 17, 1988, pp. 33–49.

Liu, D., "A case study of the HICLASS software system, a manufacturing expert system," Knowledge-Based Expert Systems for Manufacturing, Winter Annual Meeting of the American Society of Mechanical Engineers, Dec 1986, beginning on p. 25.

Liu, D., "Utilization of artificial intelligence in manufacturing," Proceedings of Autofact 6 Conference, October 1–4, 1984, pp. 2.60–2.78.

Milacic, V.R., "How to build expert systems," *Annals of the CIRP*, vol. 35, no. 2, 1986.

Mouleeswaran, C.B., "PROPLAN—a knowledge based expert system for manufacturing process planning," MS thesis, University of Chicago, Ill., 1984.

Nau, D.S. and Luce, M., "Knowledge representation and reasoning techniques for process planning: Extending SIPS to do tool selection," Proceedings of the 19th CIRP International Seminar on Manufacturing Systems, Pennsylvania State University, June 1–2, 1987, pp. 91–98.

Nau, D.S., "SIPP Reference Manual," Department of Computer Sciences, University of Maryland, 1986.

Nau, D.S., "Automated process planning using hierarchical abstraction," 1987 Texas Instruments call for papers on AI for Industrial Automation, July 1987.

Nau, D.S. and Chang, T.C., "A knowledege based approach to process planning," Bound Volume of the Symposium on Computer Aided/Intelligent Process Planning, the Winter Annual Meeting of the ASME, November 17–2, 1985, pp. 65–72.

Nau, D.S. and Chang, T.C., "Prospects for Process Selection Using AI," *Computers in Industry*, vol. 4, no. 3, 1983, pp. 253–263.

Nau, D.S. and Chang, T.C., "Prospects for process selection using artificial intelligence," *Computer-Aided Process Planning*, J. Tulkoff, (ed.), Society of Manufacturing Engineers, 1985.

Phillips, R.H., Arunthavanathan, V., and Zhou, X.D., "Symbolic representation of CAD data for artificial intelligence based process planning," PED-Vol 19, *Computer-Aided/Intelligent Process Planning* (eds Liu, C.R., Chang, T.C., and Komanduri, R.), ASME WAM, Miami Beach, Fla., November 17–22, 1985, pp. 31–42.

Ray, S., "A knowledge representation scheme for processes in an automated manufacturing environment," Proceedings of the IEEE International Conference on Systems, Man and Cybernetics, Atlanta, Ga., October 1986.

Shaw, M.J., Park, S., and Menon, U., "Incorporating machine learning in knowledge-based process planning systems: an explanation-based approach", The Beckman Institute for Advanced Science and Technology, University of Illinois at Urbana-Champaign, October 1988.

Steudel, H.J. and L.L. Firchow, "An expert system for evaluating and selecting computer-aided process planning systems," Knowledge-Based Expert Systems for Manufacturing, Winter Annual Meeting of the American Society of Mechanical Engineers, December 1986, beginning on p. 287.

Tsang, J., "The propel process planning," Proceedings of the 19th CIRP International Seminar on Manufacturing Systems, Pennsylvania State University, June 1–2, 1987, pp. 71–77.

Tsang, J.P., "Genericity in process planning systems," 1st International conference on Applications of AI to Engineering Problems, Southampton, UK, April 1986.

van't Erve, A.H. and Kals, H.J.J., "The Selection of Optimum Machining Operations in Automated Process Planning,"

van't Erve, A.H. and Kals, H.J.J., "XPLANE, a generative computer aided process planning system for part manufacturing," *Annals of CIRP*, vol. 35, no. 1, 1986, pp. 325–329.

van't Erve, A.H. and Kals, H.J.J., "The selection of optimum machining operations in automated process planning," University of Twente, Laboratory of Production Engineering, The Netherlands.

Vissa, N., A frame-based generative process planning system for machining prismatic parts, M.S. Thesis, Purdue University, December 1987.

Wang, H.-P. and Wysk, R.A., "A knowledge-based approach for automated process planning," *International Journal of Production Research*, vol. 26, no. 6, 1988. pp. 999–1014.

Wang, H.P. and Wysk, R.A., "Intelligent reasoning for process planning," Technical Paper, Department of Industrial and Systems Engineering, Penn State University, 1986.

Wang, H.P. and Wysk, R.A., "Intelligent reasoning for process planning," *Computers in Industry*, vol. 8, 1987, pp. 293–309.

Woodhead, R., Dobolyi, Z., and Pennington, A.D., "Process planning as an application for expert systems technology," Proceedings of Production Engineering Conference, ASME Winter Annual Meeting, Anaheim, Calif. pp. 143–155. 1986.

Zdeblick, W.J., "Process planning evolution: the impact of artificial intelligence," Proceedings of the 19th CIRP International Seminar on Manufacturing Systems, Pennsylvania State University, June 1–2, 1987, pp. 175–179.

A.5 Feature Recognition for Process Planning

Aldefeld, B., "On automatic recognition of 3D structures from 2-D representations," *Computer Aided Design*, vol. 15, no. 2, 1983, pp. 59–64.

Bond, A.H., Melkanoff, M.A., Ahmed, S.Z., Chang, K.J., Kim, H., and Soetarman, B., "Automatic extraction of geometric features from CAD models," Intelligent Manufacturing System, II., 2nd International Summer Seminar, (Milacic, V.R. ed.), August 24–29, 1987, pp. 143–160.

Bunce, P., Pratte, M.J., Pavey, S., and Printe, J., "Features Extraction and Process Planning," CAM-I Publication R-86-GM/PP-01, January 1986.

Butterfield, W.R., Green, M.K., Scott, D.C., and Stoker, W.J., "Part features for process planning," CAM-I Publication R-86-PPP-01, November 1986.

Chang, T.C., "Integrated process planning approach for discrete part machining," Proceedings of 14th NSF Conference on Production Research and Technology, October 6–9, 1987, pp. 43–50.

Choi, B.K. and Barash, M.M., "STOPP: an approach to CADCAM Intergration," *Computer Aided Design*, vol. 17, no. 4, May 1985, pp. 162–168.

Choi, B.K., "CAD/CAM compatible tool oriented process planning system," Ph.D. Thesis, Purdue University, 1982.

Choi, B.K., Barash, M.M., and Anderson, D.C., "Automatic recognition of machined surfaces from a 3D solid model," *Computer Aided Design*, vol. 16, no. 2, 1984, pp. 81–86.

Dong, Z. and Soom, A., "Computer-automated interpretation of 2-D CAD databases for rotational parts," PED-Vol 21, *Integrated and Intelligent Manufacturing*, (eds Liu, C.R. and Chang, T.C.), ASME, WAM, Anaheim, Calif., December 7–12, 1986, pp. 181–192.

Graves, G.R., Yelamanchili, B., and Parks, C.M., "An Interface Architecture for CAD/CAPP integration using knowledge-based systems and feature recognition algorithms," *International Journal of Computer Integrated Manufacturing*, vol. 1, no. 2, April–June 1988, pp. 89–100.

Grayer, A.R., "The automatic production of machined components starting from a stored geometric description," *Advances in Computer Aided Manufacturing*, (ed. D. McPherson), North-Holland Publishing Co., 1977, pp. 137–151.

Haralick, R.M. and Queeney, D., "Understanding engineering drawings," *Computer Graphics and Image Processing*, vol. 20, 1982, pp. 244–258.

Henderson, M.R. and Anderson, D.C., "Computer recognition and extraction of form features: a CAD/CAM link," *Computers in Industry*, vol. 5, 1984, pp. 329–339.

Henderson, M.R. and Musti, S., "Automated group technology part coding from a three-dimensional CAD database," Translations of ASME, *Journal of Engineering for Industry*, vol. 110, no. 3, August 1988, pp. 278–287.

Henderson, M.R. "Extraction of feature information from three dimensional CAD data," Ph.D. Thesis, Purdue University, 1984.

Henderson, M.R., and Chang, G.-J., "FRAPP: automated feature recognition and process planning from solid model data," Proceedings of the 1988 ASME International Computers in Engineering Conference and Exhibition, ASME, July 31–August 4, 1988, pp. 529–536.

Henderson, M.R., "Automated group technology part coding from a three-dimensional CAD database," Bound Volume, the Winter Annual Meeting of the ASME, December 1986, pp. 195–204.

Henderson, M.R., "Feature recognition in geometric modeling," CAM-I's 13th Annual Meeting and Technical Conference, November 13–15, 1894, pp. 5–1 to 5–12.

Hirschtick, J.K. and Gossard, D.C., "Geometric reasoning for design advisory systems," Proceedings of the 1986 Computers in Engineering Conference, July, 1986, pp. 263–270.

Hong, J. and Tan, X., "The similarity between shapes under affine transformation," New York University, Dept. of Computer Science, Courant Institute of Mathematical Sciences, December 1987.

Jakubowski, R. "A structural respresentation of shape and its features," *Information Sciences*, vol. 39, 1986, pp. 129–151.

Jakubowski, R. "Decomposition of complex shapes for their structural recognition," *Information Sciences*, 1989.

Jakubowski, R. "Extraction of shape features for syntactic recognition of mechanical parts," *IEEE Transactions on Systems, Man, and Cybernetics*, vol. 5, September/October 1985, pp. 642–651.

Jakubowski, R., "Syntactic characterization of machine parts shapes," *Cybernetics and Systems: An International Journal*, vol. 13, 1982, pp. 1–24.

Joshi, S. and Chang, T.C., "Feasible tool approach directions for machining holes in automated process planning systems," PED-Vol 21, *Integrated and Intelligent Manufacturing*, (eds. Liu, C.R. and Chang, T.C.), ASME, WAM, Anaheim, Calif., December 7–12, 1986, pp. 157–170.

Joshi, S. and Chang, T.C., "Graph-based heuristics for recognition or machined features from a 3D solid model," *Computer Aided Design*, March, 1987

Joshi, S., "CAD interface for automated process planning," Ph.D. Thesis, Purdue University, 1987.

Joshi, S., Vissa, N., and Chang, T.C., "Expert process planning system with solid model interface," *Expert Systems: Design and Management of Manufacturing Systems*, (Kusiak, A., ed.), Taylor and Francis, 1988, pp. 111–136.

Joshi, S., Vissa, N., and Chang, T.C., "Expert process planning system with solid model interface," *International Journal of Production Research*, vol. 26, no. 5, 1988, pp. 863–885.

Joshi, S.B. and Chang, T.C., "CAD interface for automated process planning," Proceedings of the 19th CIRP International Seminar on Manufacturing Systems, Pennsylvania State University, June 1–2, 1987, pp. 39–45.

Kakino, Y., et al., "A new method of parts description for computer-aided production planning," *Advances in Computer-Aided Manufacturing*, North-Holland, 1979.

Karinthi, R.R. and Nau, D.S., "Using algebraic properties and boolean operations to compute feature interactions," AAAI Spring Symposium on AI in Manufacturing, March 1989.

Kumar, B., Anand, D.K., and Kirk, J.A., "An intelligent feature extractor for automated machining," Proceedings of the 5th International Conference on Systems Engineering, Dayton, Ohio, September 1987.

Kumar, B., Anand, D.K., and Kirk, J.A., "Integration and testing of a intelligent feature extractor within a flexible manufacturing Protocol," Proceedings of the 16th NAMRC, Urbanan, Ill., May 1988.

Kumar, B., Anand, D.K., and Kirk, J.A., "Knowledge representation scheme for an intelligent feature extractor," Proceedings of the 1988 ASME International Computers in Engineering Conference and Exhibition, ASME, July 31–August 4, 1988, pp. 543–550.

Kung, H., "An investigation into development of process plans from solid geometric modeling representation," Ph.D. Thesis, Oklahoma State University, 1984.

Kyprianou, L. K., "Shape classification in computer aided design," Ph.D. thesis, Christ College, University of Cambridge, Cambridge, UK. 1983.

Lee, Y.C. and K.S. Fu, "Machine understanding of CSG: extraction and unification of manufacturing features," *IEEE Computer Graphics and Applications*, 1987, pp. 20–32.

Li, R.K., "A part-feature recognition system for rotational parts," *International Journal of Production Research*, vol. 26, no. 9, September 1988, pp. 1451–1475.

Liu, C.R. and Srinivasan, R., "Manufacturing process planning using syntactic pattern recognition," *Computers in Mechanical Engineering*, March, 1984, pp. 63–66.

Nnaji, B.O., "A framework for CAD-based geometric reasoning for robotic assembly language," *International Journal of Production Research*, vol. 26, no. 5, 1988, pp. 735–764.

Pratt, M.J., "Solid modeling and the interface between design and manufacture," *IEEE, Computer Graphics and Application*, July 1984, pp. 52–59.

Pratt, M.J., "Synthesis of an optimal approach to form feature modelling," Proceedings of the 1988 ASME International Computers in Engineerng Conference and Exhibition, ASME, July 31–August 4, 1988, pp. 263–274.

Ranyak, P. and Fridshal R., "CAM-I process planning features (D&T Modeler to XPS-2 Interface)," CAM-I Publication PS-87-GM-PP-01, June 1986.

Roberts, L., "Machine perception of three-dimensional solids," in J. tippett (Ed.), *Optical and Electro-Optical Information Processing*, MIT Press, Cambridge, Mass., pp. 159–197, 1965.

Shah, J., "Feature mapping and application shell," Proceedings of the 1988 ASME International Computers in Engineering Conference and Exhibition, ASME, July 31–August 4, 1988, pp. 489–496.

Shah, J.J. and Rogers, M.T., "Expert form feature modelling shell," *Computer Aided Design*, vol. 20, no. 9, November 1988, pp. 515–524.

Srinivasan, R., "A generalized analytical methodology for generative process planning," Ph.D. Thesis, Purdue University, West Lafayette, Ind., December 1986.

Srinivasan, R., Liu, C.R., and Fu, K.S., "Extraction on manufacturing details from geometric models," *Computer and Industrial Engineering*, vol. 9, no. 2, 1985, pp. 125–133.

Staley, S.M., Henderson, M.R., and Anderson, D.C., "Using syntactic pattern recognition to extract feature information from a solid modelling database," *Computers in Mechanical Engineering*, vol. 2, no. 2, 1983, pp. 61–65.

Staley, S.M., Henderson, M.R., and Anderson, D.C., "Using syntactic pattern recognition to extract feature information from a solid modelling database," *Computers in Mechanical Engineering*, vol. 2, no. 2, pp. 61–65, 1983.

van't Erve, A.H. and Kals, H.J.J., "The selection of optimum machining operations in automated process planning."

Woo, T.C., "Computer understanding of design," Ph.D. Thesis, University of Illinois, Urbana, Ill., 1975.

Woo, T.C., "Feature extraction by volume decomposition" Technical Report No., 82–4, Department of Industrial and Operations Engineering, the University of Michigan, April 1982.

Woo, T.C., "Feature extraction by volume decomposition," Conference on CAD/CAM Technology in Mechanical Engineering, M.I.T., March 1982; also, Technical Report no. 82-4, Department of Industrial Engineering, University of Michigan, Ann Arbor, Mich., April 1982.

Woodwark, J. "Shape models in computer integrated manufacture—a review," *Computer-Aided Engineering Journal*, June 1988, pp. 103–112.

Woodwark, J.R., "Some speculations on feature recognition," *Computer Aided Design*, vol. 20, no. 4, May 1988, pp. 189–196.

A.6 Automatic NC Programming in Process Planning

Armstrong, G.T., Carey, G.C., and de Pennington, A., "Numerical code generation from a geometric modeling system," *Solid Modeling by Computers from Theory to Applications*, (Pickett, M.S. and Boyse, J.W., eds.), Plenum Press, 1984, pp. 139–157.

Bard, J.F. and Feo, T.A., "The cutting path and tool selection problem in computer aided process planning," *Journal of Manufacturing Systems*, vol. 8, no. 1, 1989, pp. 17–26.

Bezier, P., *Numerical Control: Mathematics and Applications*, Wiley, 1972.

Bockholts, P.A.J.M., et al., "TNO MITURN, Programming system for lathes," Proceedings of the Second IFIP/IFAC International Conference on Programming Languages for Machine Tools, PROLAMAT '73, April 10–13, 1973.

Budde, W., "EXAPT in NC operation planning," 10th Annual Meeting and Technical Conference, NC Society, April 14–18, 1973.

Choi, B.K. and Chang, T.C., "An approach to rough cut planning in die cavity machining," Report, School of Industrial Engineering, Purdue University, 1989.

Eversheim, W. and Holz, B., "Computer aided programming of NC-machine tools by using the system AUTAP-NC," *Annals of the CIRP*, vol. 31, no. 1, 1982.

Eversheim, W., Weck, M., Scholing, H., Zuhkle, D., and Muller, W., "Off-line programming of numerically controlled industrial robots using the ROBEX-programming-systems," *Annals of CIRP*, vol. 30, no. 1, 1981, pp. 419–422.

Gould, S.S., "Surface programs for numerical control," Curved Surface in Engineering, Churchill College, March, 1972, pp. 14–18.

Grayer, A.R., "The automatic production of machined components starting from a stored geometric description," *Advances in Computer Aided Manufacturing*, (ed. McPherson, D.), North-Holland Publishing Co., 1977, pp. 137–151.

Koloc, J., "Miturn, a computer-aided production planning system for numerically controlled lathes," Proceedings of the Second International Conference on Product Development and Manufacturing Technology, University of Strathclyde, April 1971.

Kramer, T., "The design protocol, part design editor, and geometry library of the vertical workstation of the automated manufacturing research facility at the National Bureau of Standards," Document No. NBSIR 88-3731, National Bureau of Standards, 1988.

Kramer, T.R. and Jun, J.S, "Software for an automated machining work station" NBS report, July 1986.

Kramer, T.R., "Error prevention and detection in data preparation for a numerically controlled milling machine," ASME Winter Annual Meeting, 1987.

Liang, G.R. and Liu, C.R., "Automatic NC programming as a decomposable frame problem," PED-Vol 21, *Integrated and Intelligent Manufacturing*, (eds. Liu, C.R. and Chang, T.C.), ASME, WAM, Anaheim, Calif., December 7–12, 1986, pp. 171–180.

Liang, G.R. and Liu, C.R., "Logic approach to surface generation (I): single-machine problem," PED Vol 25, *Intelligent and Integrated Manufacturing Analysis and Synthesis*, Bound Volume, ASME Winter Annual Meeting, Boston, Mass., December 13–18, 1987, pp. 259–269.

Milacic, R.V. and Pilipovic, M., "Conceptual design based on the linguistic approach and the automata theory," *Annals of the CIRP*, vol. 35, no. 1, 1986.

Okawa, Y. and Miyazaki, M., "Automatic generation of NC tapes for a 3-D milling machine," Proceedings, 1985 ASME International Computers in Engineering Conference, Boston, Mass., August 4–8, 1985, vol. 1, pp. 307–315.

Persson, H., "NC machining of arbitrarily shaped pockets," *Computer Aided Design*, vol. 10, no. 3, May 1978, pp. 169–174.

Pond, J.B. "Answers to CMM problems sought after GIDEP 'Alert'," *Metalworking News*, December 1988.

Preiss, K. and Kaplansky, E., "Automated CNC milling by artificial intelligence methods," Proceedings of Autofact 6 Conference, October 1–4, 1984, pp. 2.40–2.59.

Preiss, K. and Kaplansky, E., "Automated part programming for CNC milling by artificial intelligence techniques," *Journal of Manufacturing Systems*, vol. 4, no. 1, 1985.

Stoltenkamp, H., van Houten, F.J.A.M., and Kals, H.J.J., "An interactive input-translator for the programming of turning operations," Proceedings of CAPE '83, Amsterdam, 1983.

Wang, W., "Application of solid modelling to automate machining parameters for complex parts," Proceedings of the 19th CIRP International Seminar on Manufacturing Systems, Pennsylvania State University, June 1–2, 1987.

Wong, C.L., Bagchi, A., and Ahluwalia, R.A., "DMAP: a computer integrated system for design and Manufacturing of axisymmetric parts," Bound Volume, *Knowledge-Based Expert Systems for Manufacturing*, ASME WAM, 1986, pp. 327–338.

Wong, C.L., "DMCP: a computer assisted system for the design and manufacturing of cylindrical parts," MS Thesis, the Ohio State University, 1985.

A.7 Fixturing Method Planning in Process Planning

Asada, H. and By, A., "Kinematic analysis and design for automatic workpart fixturing in flexible assembly," pp. 237–244.

Asada, H. and By, A.B., "Kinematic Analysis of Workpart Fixturing for Flexible Assembly with automatically reconfigurable fixtures," *IEEE Journal of Robotics and Automation*, vol. RA-1, no. 2, June 1985, pp. 86–93.

Asada, H. and By, A.B., "Kinematics of workpart fixturing," IEEE, 1985, pp. 337–345.

Asada, H. and West, H., "Kinematic analysis and design of tool guide mechanisms for grinding robots," *Computer-Integrated Manufacturing and Robotics*, vol. 13 (Leu, M., and Martinex, M., eds.), 1984, pp. 1–17.

Asada, H., and West, H., "Kinematic analysis and design of tool guide mechanisms for grinding robots," *Computer-Integrated Manufacturing and Robotics*, ASME WAM (Leu, M.C. and Martinez, M.R., eds.), pp. 1–17.

Bagchi, A. and Lewis, R.L., "On fixturing issues for the factory of the future," pp. 197–202.

Bazrov, B.M., "Selection of support and reference surfaces (Locations) for mounting the exchangeable elements of the MFTW System," *Soviet Engineering Research*, vol. 53, no. 5, pp. 24–27.

Bazrov, B.M., "Selection of support and reference surfaces (Locations) for mounting the exchangeable elements of the MFTW System," Stankii Instrument, *Soviet Engineering Research*, vol. 2, no. 5, 1982, pp. 94–97.

Bidanda, B. and Cohen, P.H., "An integrated CAD-CAM approach for the selection of workholding devices for concentric, rotational components," 1989 NSF Conference on Advances in Manufacturing System Integration & Process, January 1989.

Boerma, J.R., "Fixes, a system for automatic selection of set-ups and design of fixtures," University of Twente, Laboratory of Production Engineering, January 1988.

Chou, Y.C. and Barash, M.M., "A CAD system for fixture design" International Machine Tool Technology Conference, September 1982.

Chou, Y.C. and Barash, M.M., "Computerized fixture design from solid models of workpieces," PED-Vol 21, *Integrated and Intelligent Manufacturing*, (eds Liu, C.R. and Chang, T.C.), ASME, WAM, Anaheim, Calif., December 7–12, 1986, pp. 133–142.

Cohen, Paul H. and Mittal, Ravi O., A methodology for fixturing and machining of prismatic components," 1989 NSF Conference on Advances in Manufacturing System Integration & Process, January 1989.

Daimon, M., Yoshida, T., Kojima, N., Yamamoto, H., and Hoshi, T., "Study for designing fixtures considering dynamics of thin-walled plate- and box-like workpieces," pp. 319–322.

Englert, P. J. and Wright, P.K., "Principles for part setup and workholding in automated manufacturing," *Journal of Manufacturing Systems*, vol. 7, no. 2, pp. 147–161, 1988.

Ferriera, P.M., et al., "AFIX: an expert system approach to fixture design," Symposium on Computer-Aided Intelligent Process Planning, ASME Winter Annual Meeting, (Liu, C.R., Chang, T.C. and Komanduri, R. eds.) 1985, pp. 73–82.

Gandhi, M.V. and Thompson, B.S., "Flexible fixturing based on the concept of material phase-change," Proceedings, CAD/CAM, Robotics and Automation International Conference, February 13–15, 1985, pp. 471–474.

Gandhi, M.V. and Thompson, B.S., "Phase change fixturing for flexible manufacturing Systems," *Journal of Manufacturing Systems*, vol. 4, no. 1, 1985, pp. 29–38.

Gandhi, M.V. and Thompson, B.S., "The integration of CAD and

CAM in adaptive fixturing for flexible manufacturing systems," pp. 301–305.

Gripo, P.M., Gandhi, M.V., and Thompson, B.S., "The computer-aided design of modular fixturing systems," the *International Journal of Advanced Manufacturing Technology*, 1987, pp. 75–88.

Hayashi, A., Yamazaki, K., Hoshi, T., Misawa, K., and Suzuki, N., "Development of fixture design and fabrication system by means of CAD/CAM," 16 North American Manufacturing Research Conference Proceedings, NAMRC May 24–27, 1988, pp. 429–436.

Ingrand, F. and Latombe, J., "Functional reasoning for automatic fixture design," CAM-I's 13th Annual Meeting & Technical Conference, November 13–15, 1984, pp. 8-53–8-65.

Jiang, W., Wang, Z., and Cai, Y., "Computer-aided group fixture design," January, 1988, pp. 145–148.

Joneja, A. and Chang, T.C., "a generalized framework for automatic planning of fixture configuration," ASME Winter Annual Meeting, December 10–15, 1989.

Lee, J.D. and Haynes, L.S., "Finite-element analysis of flexible fixturing system," Transactions of the ASME, *Engineering for Industry*, vol. 109, May 1987, pp. 134–139.

Mani, M. and Wilson, W.R.D., "Automated design of workholding fixtures using kinematic contraint synthesis," 16th North American Manufacturing Research Conference Proceedings, Fixture-Design May 24–27, 1988. pp. 437–444.

Mani, M., and Wilson, R.D., "Avoiding interence in 2D fixture and grasp planning," Proceedings of 1988 ASME International Computers in Engineering Conference, July 31–August 4, 1988, pp. 397–402.

Markul, A., Markusz, Z., Farkas, J., and Filemon, J., "Fixture design using prolog: an expert system," *Robotics & Computer-Integrated Manufacturing*, vol. 1, no. 2, 1984, pp. 167–172.

Nee, A.Y.C., Bhattacharyya, N., and Poo, A.N., "Applying AI in jigs and fixtures design," *Robotics & Computer-Integrated Manufacturing*, vol. 3, no. 2, 1987, pp. 195–200.

Nguyen, V.-D., "The synthesis of stable grasps in the plane," Proceedings of the 1986 IEEE Conference on Robotics & Automation, vol. 2 April 7–10, 1986, pp. 884–889.

Shah, J., "CLAMPS—automated fixturing in a flexible manufacturing environment," M.S. thesis, School of Industrial Engineering, Purdue University, 1988.

Youcef-Toumi, K., Liu, W.S., and Asada, H., "Computer-aided analysis of reconfigurable fixtures and sheet metal parts for robotic drilling," *Robotics & Computer Integrated Manufacturing*, vol. 4, no. 3/4, 1988, pp. 387–393.

Index

D

E

F

G

H

I

J

K

L